T0281368

Lecture Notes in Computer Science 10099

More information about this series at http://www.springer.com/series/7408

Andrea Bondavalli · Sara Bouchenak
Hermann Kopetz (Eds.)

Cyber-Physical Systems of Systems

Foundations – A Conceptual Model and Some Derivations: The AMADEOS Legacy

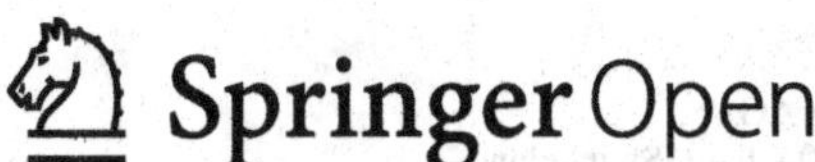

Editors
Andrea Bondavalli
Department of Mathematics and Informatics
University of Florence
Florence
Italy

Hermann Kopetz
Institute of Computer Engineering
Vienna University of Technology
Vienna
Austria

Sara Bouchenak
INSA Lyon - LIRIS
Lyon
France

ISSN 0302-9743 ISSN 1611-3349 (electronic)
Lecture Notes in Computer Science
ISBN 978-3-319-47589-9 ISBN 978-3-319-47590-5 (eBook)
DOI 10.1007/978-3-319-47590-5

Library of Congress Control Number: 2016958992

LNCS Sublibrary: SL2 – Programming and Software Engineering

Printed on acid-free paper

This Springer imprint is published by Springer Nature
The registered company is Springer International Publishing AG
The registered company address is: Gewerbestrasse 11, 6330 Cham, Switzerland

Foreword

"We are called to be architects of the future, not its victims"
Richard Buckminster-Fuller

This book is about "systems-of-systems". If you search in Google for this term, you get *176,000,000 results in 0.60 seconds* (20.9.2016). This clearly shows the importance and vibrancy of this field! However, it also shows the wide and diverging variety of viewpoints, concepts, and opinions related to systems-of-systems.

Technical systems-of-systems – in the form of networked, independent constituent computing systems temporarily collaborating to achieve a well-defined objective – form the backbone of most of today's infrastructure. The energy grid, most transportation systems, the global banking industry, the water-supply system, military equipment, many embedded systems, and a great number more, strongly depend on systems-of-systems. The correct operation and continuous availability of these underlying systems-of-systems are fundamental for the functioning of our modern society.

Looking at such systems-of-systems, one property clearly stands out: *complexity*. Modern systems-of-systems have reached a degree of structural and behavioral complexity that makes it difficult – in many cases impossible – to understand them. As a consequence, a massive engineering effort and monetary investment are required to design, implement, maintain and evolve many of today's systems-of-systems. Owing to new properties that are introduced when systems-of-systems are formed – such as emergent behavior, especially *unpredictable* emergent behavior – a new element of *risk* is also introduced. Because our dependence on these growing systems-of-systems is nearly total, we need reliable methods, principles, and tools to manage the evolution of our systems-of-systems in today's world of growing complexity, relentless change, and merciless uncertainty. This book is a move forward on this interesting and important path.

The first important step of achieving this objective is the development of an understandable and consistent *set of concepts* describing the systems-of-systems domain. This is not the case in the current state of the art: Therefore, this is the first valuable contribution of this book for the community.

Systems-of-systems become alive by exchanging information and control between their constituent systems and the physical environment via *interfaces*. Interfaces are responsible for many properties in systems-of-systems and therefore need detailed attention: This is the second impressive result of the book – a thorough treatment of interface definition, specification, implementation, and monitoring.

The most fascinating and disturbing phenomenon in systems-of-systems is *emergence*: Behavior or properties that only become active or visible when the constituent systems start cooperating. Emergence has been studied in many contexts and with many objectives: Here we find a consistent theory with important novel concepts, which is applicable to many systems. This is a major research achievement.

Next, a rich *conceptual model* of generic systems-of-systems, divided into ten viewpoints, is developed. The conceptual model is supported by a SySML profile, covering the ten viewpoints. Especially interesting and innovative parts are the representation and use of *time* in the SoS time package and the handling of *emergence* in the SoS emergence package. As additional material, a three-level architecture description framework for generic systems-of-systems – again well implementing the ten viewpoints – is presented, which can be used in a commercially available graphical tool. This part greatly helps the understanding of systems-of-systems development and documentation.

One of the very strong points of this book is the presentation of time and *synchronized timing* in systems-of-systems. This aspect has not been covered with sufficient theoretical rigor in the existing literature. Topics are global time base and resilient clocks – presenting both innovative research and an excellent tutorial.

Many systems-of-systems are not static, but must adapt to changing requirements, be it changing business requirements or in response to changing environmental or operational parameters of the constituent systems. Timely adaptation of systems-of-systems requires a property that is called *dynamicity*. A theory – based on autonomic computing – and implementation patterns for coping with dynamicity are presented.

This volume is on a research and technology level and throughout the book, interesting and illustrative examples can be found.

The book grew out of a sequence of EU-funded projects of which the described project AMADEOS was the culmination. Most of the book shows an admirable maturity, both of the material and the presentation, and is certainly a source of much more fruitful research.

I wish the reader as much satisfaction in reading this rewarding book as the authors had during writing it!

September 2016 Frank J. Furrer

Preface

The general availability of a powerful communication infrastructure (e.g., the Internet) makes it possible to interconnect existing self-contained computer-systems— called *Constituent Systems (CS)*— that already provide a useful service to their users. Such a composition of a set of independent and autonomous systems that brings about novel services to their stakeholders is called a *System-of-Systems* (SoS). The purpose of building an SoS out of CSs is to realize *emergent services* that go beyond the services provided by any of the isolated CSs. Emergence is thus at the core of SoS engineering.

Consider, for example, a local bank terminal that is connected to the worldwide ATM system. This connection enables the novel service of worldwide accessibility of a local bank account. A tourist in a remote country can withdraw money, denoted in the currency of the host country, from his/her home bank that displays the transaction in the currency of the home bank. Since the exchange rate of the currencies is time dependent, the monetary value of this multi-currency transaction depends on the instant when the transaction is executed. This simple example shows that new issues, such as the appropriate representation of information in differing contexts or the time when an action takes place, play an important role in such an SoS.

The vision of the Internet of Things (IoT) assumes that the networked connection of *smart things*, i.e., *Cyber-Physical System (CPS)* with sensors and actuators that observe and directly influence the physical environment, has the potential to provide disruptive novel services to our society. We call such an integration of stand-alone CPSs that provides services that go beyond the services of any of its isolated CPSs a *Cyber-Physical System of Systems (CPSoS)*.

Consider, for example, a smart grid where a multitude of autonomous energy producers and energy consumers, controlled by their local CPSs and the control systems of the utility companies cooperate to provide a smooth flow of electric energy from producers to consumers. Despite the fact that such a possibly gigantic CPSoS is in a continuous state of evolution, it must provide a dependable service 24 hours a day, seven days a week.

This book on *Systems of Systems* documents the main insights on CPSoS that we gained during our work in the European research project AMADEOS (acronym for Architecture for Multi-criticality Agile Dependable Evolutionary Open System-of-Systems). The objective of this research was to bring *time awareness, dynamicity,* and *evolution* into the design of SoS, to establish a sound *conceptual model* that provides a well-defined language for describing SoS, to investigate the intricate topic of *emergence* in an SoS, and to outline a *generic architectural framework* and an *SoS design methodology*, supported by some prototype tools, for the modeling, development, and evolution of time-sensitive SoS.

The AMADEOS partners from industry (Thales Nederlands, Resiltech, and ENCS) and universities (University of Firenze, Vienna University of Technology and Université Grenoble Alpes) joined efforts to arrive at results that are, on the one hand, of wide-ranging value in an industrial context, and on the other hand, extend the understanding of SoS beyond the current state of the art in the academic world.

It is the objective of this book to present in a single consistent body the foundational concepts and their relationships in order to form a conceptual basis for the description and understanding of SoS and to go deeper in what we consider the characterizing and distinguishing elements of SoS: time, emergence, evolution, and dynamicity.

The first part of the book is devoted to this conceptual work. We start in Chap. 1 with the set of definitions of the relevant concepts. The need for a new approach and vision of interfaces is the topic of Chap. 2. In Chap. 3 we investigate the phenomenon of emergence in CPSoS, with a definition of emergence in the SoS context, and discuss some properties of emergent phenomena. Chap. 4 provides the definition of the AMADEOS conceptual model that captures SoS basic concepts and their interrelationships, and describes a SysML profile semi-formalization supporting the definition of SoS platform independent models (PIMs).

Part 2 of the book deals with the engineering framework developed in AMADEOS and the technical solutions adopted to deal with time, dynamicity, and evolution. More precisely, Chap. 5 defines the overall tool-supported AMADEOS Architectural Framework (AF), with its main building blocks and interfaces. Chap. 6 elaborates on the role of time and clocks in SoS and presents the design of a Resilient Master Clock (RMC), a hardware–software solution that has been developed in the course of the AMADEOS project. The final chapter in this part presents the AMADEOS dynamicity management and the different Monitor–Analyze–Plan–Execute (MAPE) components.

Part 3 of the book contains case studies of smart grid applications to demonstrate the suitability of the AMADEOS methodology for the design of an advanced industrial SoS.

In the following, we present a short overview of the main points covered in each one of the chapters of this book.

The objective of Chap. 1 is the development of a set of coherent concepts and the associated terms that can be used by domain experts to communicate their ideas about SoS. We start form the fundamental notion of a *Constituent System (CS)* that is time aware and consider an SoS as an integration of a finite number of CSs that are independent and operable, and which are networked together for a period of time to achieve a certain higher goal. The CSs interact by the timely exchange of information items (we call them Itoms) across *Relied Upon Interfaces* (RUI). An Itom is an atomic triple of data, explanation of the data and time. It follows a detailed model of time, based on Newtonian physics, and time measurement by digital clocks is presented. This model of time is used to define the notion of the state of a system at a given instant as the totality of the information items from the past that can have an influence on the future behavior of a system. We then discuss the characteristics of three basic communication mechanisms in cyberspace: datagrams, event-triggered positive acknowledgment, and retransmission protocols, as well as time-triggered protocols, followed by an elaboration of the information flow across stigmergic channels in the physical environment. After a short passage on interfaces (which are discussed at length in Chap. 2) the concepts of *dynamicity* and *evolution* are treated in the last section of Chap. 1.

The focus of Chap. 2 is on the important role that interfaces play in the control of the cognitive complexity of the models that explain the behavior of an SoS. The boundaries among the CSs within an SoS are formed by the RUI of the CSs. The precise specifications of the syntactic, semantic, and temporal properties of these RUIs hide the internals of a CS implementation and are a good example for the application of the well-known divide and conquer principle. Interface layers allow for the discussion of system interface properties at different abstraction levels. Three interface layers are introduced: the *cyber-physical layer*, the *informational layer*, and the *service layer*. The cyber-physical layer is concerned with the reliable transport of context-sensitive bit-patterns across RUIs, both via messages in cyber-space and stigmergic channels in the physical environment. The informational layer abstracts from the context-sensitivity and the concrete technical implementations of the cyber-physical layer. The service layer structures the behavior of a system into a set of capabilities enabling management of dynamicity and evolution at the interface level. The specification of the execution semantics of an RUI assumes a frame-based synchronous data flow model. In many SoS the connections between the RUIs of the CSs are not static, but dynamic. The sections in Chap. 2 on dynamicity and managed evolution give details on the reaction and reconfiguration capabilities of an SoS that can be considered in the CPSoS design at the interface level.

Chap. 3 deals with the important topic of *emergence* in CPSoS. As quoted earlier, emergence is at the core of SoS engineering. The essence of the concept of emergence is aptly communicated by the following quote, attributed to Aristotle: "The whole is greater than the sum of its parts." The interactions of *parts* (the CSs) can generate a *whole* (the SoS) with unprecedented properties that go beyond the properties of any of its constituent parts. The immense varieties of inanimate and living entities that are found in our world are the result of emergent phenomena that have a small number of elementary particles at their base. After a lengthy discussion about the importance of multi-level hierarchies in the models of nearly decomposable complex systems, the following definition of emergence is presented: *A phenomenon of a whole at the macro-level is emergent if and only if it is of a new kind with respect to the non-relational phenomena of any of its proper parts at the micro level.* In the following sections of Chap. 3 the concepts of *downward causation* and *supervenience* are explained and it is conjectured that in a multi-level hierarchy emergent phenomena are likely to appear at the macro-level when there is a causal loop formed between the micro-level that forms the whole and the whole (i.e., the ensemble of parts) that constrains the behavior of the parts at the micro-level. A schema for the classification of emergent phenomena is presented and four concrete examples of emergent phenomena in computer systems are given. In the final section of Chap. 3, the focus is on the analysis of *detrimental emergent phenomena* in safety-critical CPSoS.

Chap. 4 covers the AMADEOS SysML profile for SoS conceptual modeling. The focus is on the definition of a SysML profile as a modeling support for representing the AMADEOS SoS conceptual model. The basic SoS concepts and their relationships are modeled using a SysML semi-formal representation according to different viewpoints, which represent the key perspectives of AMADEOS: structure, dynamicity and evolution, dependability and security, time, multi-criticality and emergence. Finally, a Smart Grid household scenario is introduced to exemplify the application of the profile and to instantiate the basic SoS concepts to a concrete case study from the Smart Grid domain, focusing on the architecture and emergence viewpoints.

Chap. 5 introduces the overall tool-supported *AMADEOS Architectural Framework (AF)* with its main building blocks and interfaces. The high-level representation of the AMADEOS AF is shown as a pyramid made of four different layers, namely, mission, conceptual, logical, and implementation. Apart from the mission block, all the remaining levels are organized in slices, each corresponding to a specific viewpoint. The following viewpoints of an SoS are explored: structure, dependability, security, emergence, and multi-criticality. Finally, for SoS modeling, a supporting facility tool based on Blockly is demonstrated. Blockly is a visual Domain-Specific Language (DSL) and has been adopted to ease the design of SoS by means of a simple and intuitive user interface; thus requiring minimal technology expertise and support for the SoS designer.

Chap. 6 stipulates that a global notion of time with known precision, shared by all CSs, is essential for the dependable operation of an SoS. Such a global notion of time is needed to specify the temporal properties of interfaces, to enable the interpretation of timestamps in the different CSs, to limit the validity of real-time data, to synchronize input and output actions across CSs, to provide conflict-free resource allocation, to perform prompt error detection, and to strengthen security protocols. Since CSs can join and leave the SoS dynamically, external clock synchronization is the preferred alternative in an SoS. Such an external clock synchronization can be based on the standardized time signal distributed worldwide by Global Navigation Satellite Systems (GNSS), such as GPS, Galileo or GLONASS. Since a GNSS time signal can become unavailable, a resilient master clock is proposed to extend the holdover interval after a time-signal failure. In AMADEOS a prototype of such a resilient GPS disciplined master clock has been developed and tested. The chapter describes the design, implementation and validation of this resilient master clock prototype.

Chap. 7 is devoted to the management of dynamicity in SoS. The well-known *Monitor–Analyze–Plan–Execute (MAPE)* control loop, developed by IBM in the context of autonomic computing, provides the framework for the management of dynamicity. When a Service Level Agreement (SLA) and its associated Service Level Objectives (SLOs) are associated with the service of a managed element, the MAPE control loop guarantees that these SLOs are met. If this is not the case, a new plan is calculated and used to reconfigure the system. This chapter presents AMADEOS dynamicity management and the components of MAPE.

Finally, Chap. 8 contains three case studies from the smart grid domain to demonstrate the viability of the AMADEOS approach to the design of SoS. The three case studies, electric vehicle charging, household management, and an integrated case study that combines the first two together with ancillary services, are modeled by using the AMADEOS Architectural Framework (AF) and the AMADEOS tool set. We utilize the four levels of the AMADEOS AF – mission, conceptual, logical, and implementation – as well as the seven viewpoints that have been defined: structure, dynamicity, evolution, dependability and security, time, multi-criticality, and emergence.

Modeling complex and pervasive infrastructures like the one used as case study clearly highlights how the support of a precise conceptual model and of specific tools for its instantiation is fundamental for a sound and comprehensive codification of the various properties of the whole. At design time the identification of causal loops in the lower levels of the hierarchy, enabled by the support for simulation through model execution, is a mandatory step to identify possible emergent behaviors at the higher levels. In fact, such behaviors may lead, also in the future evolution of the SoS, to a violation of system requirements. A correct representation of the environment has also proven to be necessary. Finally, global time awareness and monitoring are fundamental for the early detection and for containing the effect of detrimental emergence phenomena at run time.

Although the chapters of the book are arranged in a logical order, an effort has been made to keep each chapter self-contained. The book contains also a glossary of all the terms and concepts used to ease reading and provide a reference for relevant terms in the domain of SoS.

This book can be used as a textbook or supplemental reading for advanced teaching on SoS, their concepts, and their design. In addition to this book, a set of slides is available that helps a lecturer in the development of the teaching material for an advanced course on Systems of Systems, (http://rcl.dsi.unifi.it/projects/amadeos/amadeosteachingmaterial).

Finally, we would like to thank all the following experts for reviewing and helping to improve this book: Wilfried Elmenreich (Alpen-Adria-Universität Klagenfurt, Austria), and Wilfried Steiner (TTTech, Austria), Lorenzo Falai (Resiltech), Leonardo Montecchi (University of Florence), and Antoine Boutet (LIRIS).

Hermann Kopetz
Andrea Bondavalli
Sara Bouchenak

Acknowledgments

This work was partially supported by the FP7-610535-AMADEOS project.

Contents

Basic Concepts on Systems of Systems

Andrea Ceccarelli[1(✉)], Andrea Bondavalli[1], Bernhard Froemel[2], Oliver Hoeftberger[2], and Hermann Kopetz[2]

[1] Department of Mathematics and Informatics, University of Florence, Florence, Italy
{andrea.ceccarelli,andrea.bondavalli}@unifi.it
[2] Institute of Computer Engineering, Vienna University of Technology, Vienna, Austria
{froemel,oliver}@vmars.tuwien.ac.at, h.kopetz@gmail.com

1 Introduction

A System of System (SoS) stems from the integration of existing systems (legacy systems), normally operated by different organizations, and new systems that have been designed to take advantage of this integration. Many of the established assumptions in classical system design, such as e.g., the *scope of the system is known*, that *the design phase of a system is terminated by an acceptance test* or *that faults are exceptional events,* are not justified in an SoS environment. This is well represented by Table 1.

In this chapter we present the fundamental concepts for Systems of Systems engineering established within the AMADEOS[1] project, with the objective of proposing a shared System of Systems *vocabulary* and define an *implicit theory* about the SoS domain.

The overarching concern of our work is to target the reduction of the *cognitive complexity* needed to comprehend the behaviour of a SoS by the application of appropriate *simplification strategies* [29]. In fact, the considerable cognitive effort needed to understand the operation of a large SoS is the main cause for the substantial engineering (and monetary) effort required to design and maintain many of today's Systems of Systems.

Our position is that the first important step of achieving simplicity is the development of an understandable set of concepts that describes the SoS domain. In fact, if the language used to talk about a domain contains terms with clearly defined meanings that are shared by a wide part of the domain community, then the communication of ideas among experts is simplified. Consequently, starting from a detailed analysis of the existing concepts for the SoS domain (e.g., from the projects DANSE [35], DSoS [7],

This work has been partially supported by the FP7-610535-AMADEOS project.

[1] FP7-ICT-2013-10-610535 AMADEOS: Architecture for Multi-criticality Agile Dependable Evolutionary Open System-of-Systems, http://amadeos-project.eu/.

A. Bondavalli et al. (Eds.): Cyber-Physical Systems of Systems, LNCS 10099, pp. 1–39, 2016.
DOI: 10.1007/978-3-319-47590-5_1

Table 1. Comparison of an SoS compared to a monolithic system [22]

Characteristic	Monolithic system	System-of-system
Scope of the System	Fixed (known)	Not known
Clock Synchronization	Internal	External (e.g., GPS)
Structure	Hierarchical	Networked
Requirements and Spec.	Fixed	Changing
Evolution	Version Control	Uncoordinated
Testing	Test Phases	Continuous
Implementation Technology	Given and Fixed	Unknown
Faults (Physical, Design)	Exceptional	Normal
Control	Central	Autonomous
Emergence	Insignificant	Important
System Development	Process Model	???

COMPASS [36]), in this work we reuse main existing concepts, and formulate new ones when needed, to propose a *shared vocabulary for Systems of Systems* that can be used in a coherent and consistent manner.

The concepts devised are divided in ten viewpoints, summarized below. Noteworthy, an extended version of the conceptual model is freely available at [33].

Fundamental System Concepts. Discussed in Sect. 2, the first group focuses on the static structure of systems and their parts. It starts from the presentation of the universe and time of discourse of an SoS, to finally define an SoS and its related parts.

Time. Discussed in Sect. 3, this group explains the progression of time and its role in an SoS. The role of time and clocks in SoSes is further debated in Chap. 6.

Data and state. Discussed in Sect. 4, this groups defines the data and information that are exchanged between the Constituent Systems that form an SoS. These concepts are further investigated in Chap. 2.

Actions and Behaviour. Discussed in Sect. 5, this group illustrates the dynamics of an SoS, that consists of discrete variables by an *event-based view* or by a *state-based view*.

Communications. Discussed in Sect. 6, the focus of this group is on the role and properties of a communication system in an SoS. These concepts are further elaborated in Chap. 2.

Interfaces. Discussed in Sect. 7, this group presents the fundamentals definition for the interfaces i.e., the points of interaction of SoS components with each other and with the environment over time. These concepts are further debated in Chap. 2.

Evolution and Dynamicity. Discussed in Sect. 8, this group explains SoS dynamicity, intended as short term changes, and evolution, intended as long term changes. These concepts are also largely applied in Chap. 7.

System design and tool. Discussed in Sect. 9, this group sets the foundational concepts to define design methodologies to engineer SoSes.

Dependability and Security. Discussed in Sect. 10, this group presents dependability and security concepts, in compliance with the taxonomies presented in [1, 3, 4].

Emergence. The phenomenon of emergence in Cyber-Physical Systems of Systems and its main concepts are largely discussed in Chap. 3. Consequently, this group of concepts is only briefly introduced in this Chapter. The definition of emergence reported in Chap. 3 states that "a phenomenon of a whole at the macro-level is emergent if and only if it is of a new kind with respect to the non-relational phenomena of any of its proper parts at the micro level". Emergent phenomena can be of a different nature either beneficial or detrimental and either expected or unexpected. Managing emergence is essential to avoid undesired, possibly unexpected situations generated from CSs interactions and to realize desired emergent phenomena being usually the higher goal of an SoS [30]. For example, system safety has been acknowledged as an emerging property [31], because its meaning at the SoS level does not have the same meaning for the individual CS, and obviously it cannot be expressed just as the sum of the individual parts. Additionally, to further strengthen on the relevance of emergence in an SoS, we remark that it is acknowledged that the SoS may be exposed to new security threats when novel phenomena arise [32, 34].

2 Fundamental System Concepts

2.1 Universe and Time of Discourse

We start by delineating the universe and time of discourse of an *SoS*.

Universe of Discourse (UoD): The Universe of Discourse comprises the set of entities and the relations among the entities that are of interest when modeling the selected view of the world.

The word *domain* is often used as synonym to the notion *Universe of Discourse.*

Interval of Discourse (IoD): The Interval of Discourse specifies the time interval that is of interest when dealing with the selected view of the world.

In order to structure the *UoD* during the *IoD*, we must identify objects that have a distinct and self-contained existence.

Entity: Something that exists as a distinct and self-contained unit.

We distinguish between two very different kinds of entities, *things* and *constructs*.

Thing: A physical entity that has an identifiable existence in the physical world.

Referring to the *Three World Model of Popper* [8] there is another class of entities, we call them *constructs* that have no physical existence on their own but are products of the human mind.

Construct: A non-physical entity, a product of the human mind, such as an idea.

2.2 Systems

We use the definition of the term *system* introduced in the EU Project DSOS (Dependable System-of-systems IST-1999-11585 [9]):

System: An entity that is capable of interacting with its environment and may be sensitive to the progression of time.

By 'sensitive to the progression of time' we mean the system may react differently, at different points in time, to the same pattern of input activity, and this difference is due to the progression of time. A simple example is a time-controlled heating system, where the temperature set-point depends on the current time [9]. The role of humans in a system is discussed at length below in this section.

Environment of a System: The entities and their actions in the UoD that are not part of a system but have the capability to interact with the system.

In *classical system engineering*, the first step in the analysis of a system is the establishment of an *exact boundary* between the system under investigation and its environment.

System Boundary: A dividing line between two systems or between a system and its environment.

In SoS Engineering, such an approach can be problematic, because in many SoS the system boundary is dynamic. Consider, e.g., a *car-to-car SoS* that consists of a *plurality of cars* cruising in an area. Where is the boundary of such an SoS? A good concept should be stable, i.e., its important properties, such as size, should remain fairly constant during the IoD. The *boundary* of the *car-to-car SoS* does not satisfy this requirement and is thus a *poor concept*. Our analysis of many other existing SoSs, e.g., the worldwide ATM system or a smart grid system came to a similar conclusion: it is hardly possible to define a stable boundary of an SoS [22, 29].

In the above example of a *car-to-car SoS* each individual car in the system (consisting of the *mechanics of the car*, the *control system within the car* and the *driver*) can be considered as an *autonomous system* that tries to achieve its given objective without any control by another system.

Autonomous System: A system that can provide its services without guidance by another system.

Before starting with the detailed design of a large system an overall blueprint that establishes the framework of the evolving artifact should be developed.

System Architecture: The blueprint of a design that establishes the overall structure, the major building blocks and the interactions among these major building blocks and the environment.

Every organization that develops a system follows a set of explicit or implicit rules and conventions, e.g., naming conventions, representation of data (e.g., *endianness* of data), protocols etc. when designing the system. This set of explicit or implicit rules and conventions is called the *architectural style.*

Many of the existing legacy systems have been designed in the context of a single organization that follows its often ill-documented idiosyncratic architectural style. For example, undocumented implicit assumptions about the attributes of data can lead to mismatches when data is sent from one subsystem to another subsystem in an SoS.

Monolithic System: A system is called monolithic if distinguishable services are not clearly separated in the implementation but are interwoven.

Many systems are not monolithic wholes without any internal structure, but *are composed of interrelated parts, each of the latter being in turn hierarchic in structure until we reach some lowest level of elementary subsystem [11], p. 184.*

Subsystem: A subordinate system that is a part of an encompassing system.

We call the subsystems of a System of Systems (SoS) *Constituent Systems (CSs).*

Constituent System (CS): An autonomous subsystem of an SoS, consisting of computer systems and possibly of controlled objects and/or human role players that interact to provide a given service.

The decomposition of a system into subsystems can be carried out until the internal structure of a subsystem is of no further interest. The systems that form the lowest level of a considered hierarchy are called *components.*

Some systems can be decomposed without loss into well-understood parts, so that the functions at the system level can be derived from the functions of the parts [25].

Cyber-Physical System (CPS): A system consisting of a computer system (the cyber system), a controlled object (a physical system) and possibly of interacting humans.

An *interacting human* can be a prime mover or role player.

Prime mover: A human that interacts with the system according to his/her own goal.

An example for a prime mover could be a legitimate user or a malicious user that uses the system for his/her own advantage. A human who is a prime mover can be considered to be a constituent system (CS).

Role player: A human that acts according to a given script during the execution of a system and could be replaced in principle by a cyber-physical system.

A CPS is composed not only of the computer system, i.e., the cyber system, but also of a controlled object and possibly a human role player.

Entourage of a CPS: The entourage is composed of those entities of a CPS (e.g., the role playing human, controlled object) that are external to the cyber system of the CPS but are considered an integral part of the CPS.

2.3 System-of-Systems

We decided to select the following definition of Jamishidi as the starting point of our work [10].

System-of-Systems (SoS): An SoS is an integration of a finite number of constituent systems (CS) which are independent and operable, and which are networked together for a period of time to achieve a certain higher goal.

We consider the phrase *that are networked together for a period of time* an important part of this definition, since it denotes that a static scope of an SoS may not exist and the boundary between an SoS and its environment can be dynamic.

Dahmann and Baldwin have introduced the following four categories of SoSs [11]:

Directed SoS: An SoS with a central managed purpose and central ownership of all CSs. An example would be the set of control systems in an unmanned rocket.

Acknowledged SoS: Independent ownership of the CSs, but cooperative agreements among the owners to an aligned purpose.

Collaborative SoS: Voluntary interactions of independent CSs to achieve a goal that is beneficial to the individual CS.

Virtual SoS: Lack of central purpose and central alignment.

While a *directed SoS,* e.g., the CSs in an automobile that are under strict central management and ownership of a car company, comes close to a homogenous system, the other extreme, a *virtual CS*, lacks the elements of homogeneity and is formed by heterogeneous subsystems belonging to very different organizations.

We call an interface of a CS where the services of a CS are offered to other CSs a *Relied Upon Interface (RUI). It is '"relied upon" with respect to the SoS,* since the service of the SoS as a whole relies on the services provided by the respective CSs across the RUIs.

Relied upon Interface (RUI): An interface of a CS where the services of the CS are offered to other CSs.

In addition to a *Relied upon Message Interface (RUMI)* where messages containing information are exchanged among the CSs, a *Relied upon Physical Interface (RUPI)* where things or energy are exchanged among the CSs can exist.

Relied upon Message Interface (RUMI): A message interface where the services of a CS are offered to the other CSs of an SoS.

Relied upon Physical Interface (RUPI): A physical interface where things or energy are exchanged among the CSs of an SoS.

Relied upon Service (RUS): (Part of) a Constituent System (CS) service that is offered at the Relied Upon Interface (RUI) of a service providing CS under a Service Level Agreement (SLA).

There may be other interfaces to systems external to a CS which we investigate in Sect. 6, together with the issues of *information exchange* and *interface specification.*

3 Time

The focus of the previous Section was on the *static structure* of systems and their parts. In this Section we start being concerned with *change*. The concept of *change* depends on the progression of *time* that is one of the core topics that is investigated in AMADEOS. In an SoS a *global notion of time* is required in order to

- Enable the interpretation of timestamps in the different CSs.
- Limit the validity of real-time control data.
- Synchronize input and output actions across nodes.
- Provide conflict-free resource allocation.
- Perform prompt error detection.
- Strengthen security protocols.

We base our model of time on *Newtonian physics* and consider time as an independent variable that progresses on a dense time-line from the past into the future. For a deep discussion on the issues of time we refer to the excellent book by Withrow, *The Natural Philosophy of Time* [12] that considers also the revision to the Newtonian model by the theory of relativity. From the perspective of an SoS the *relativistic model of time* does not bring any new insights above those of the Newtonian model of time.

3.1 Basics on Time

Time: A continuous measurable physical quantity in which events occur in a sequence proceeding from the past to the present to the future.

This definition of *time,* which denotes *physical time,* has been adapted from *Dictionary.com* and uses a number of fundamental concepts that cannot be defined without circularity.

Timeline: A dense line denoting the independent progression of time from the past to the future.

The directed time-line is often called the *arrow of time*. According to Newton, time progresses in dense (infinitesimal) fractions along the arrow of time from the past to the future.

Instant: A cut of the timeline.

Event: A happening at an instant.

An event is a happening that reports about some change of state at an instant.

Signal: An event that is used to convey information typically by prearrangement between the parties concerned.

Instants are totally ordered, while events are only partially ordered. More than one event can happen at the same instant.

Temporal order: The temporal order of events is the order of events on the timeline.

Causal order: A causal order among a set of events is an order that reflects the cause-effect relationships among the events.

Temporal order and causal order are related, but not identical. Temporal order of a cause event followed by an effect event is a necessary prerequisite of causal order, but causal order is more than temporal order [2], p. 53.

Interval: A section of the timeline between two instants.

While an interval denotes a section of the timeline between two instants, the duration informs about the length only, but not about the position of such a section.

Duration: The length of an interval.

The *length of an interval* can only be measured if a standard for the duration is available. The *physical SI second* is such an international standard (the *International System of Units* is abbreviated by *SI*).

Second: An internationally standardized time measurement unit where the duration of a second is defined as 9 192 631 770 periods of oscillation of a specified transition of the Cesium 133 atom.

The physical second is the same in all three important universal time standards, UTC, TAI and GPS time. UTC (*Universal Time Coordinated*) is an astronomical time standard that is aligned with the rotation of the earth. Since the rotational speed of the earth is not constant, it has been decided to base the SI second on atomic processes establishing the International Atomic Time *TAI (Temps Atomique International).* On January 1, 1958 at 00:00:00 TAI and UTC had the same value. The TAI standard is chronoscopic and maintained as the weighted average of the time kept by over 200 atomic clocks in over 50 national laboratories. TAI is distributed world-wide by the satellite navigation system GPS (*Global Positioning System*).

Offset of events: The offset of two events denotes the duration between two events and the position of the second event with respect to the first event on the timeline.

The position of an instant on a standardized timeline can only be specified if a starting point, the origin, for measuring the progression of time (in seconds) has been established.

Epoch: An instant on the timeline chosen as the origin for time-measurement.

GPS represents the progression of TAI time in weeks and full seconds within a week. The week count is restarted every 1024 weeks, i.e., after 19.6 years. The Epoch of the GPS signal started at 00:00:19 TAI on January 6, 1980 and again, after 1024 weeks at 00:00:19 TAI on August 22, 1999.

Cycle: A temporal sequence of significant events that, upon completion, arrives at a final state that is related to the initial state, from which the temporal sequence of significant events can be started again.

An example for a cycle is the rotation of a crankshaft in an automotive engine. Although the duration of the cycle changes, the sequence of the significant events during a cycle is always the same.

Period: A cycle marked by a constant duration between the related states at the start and the end of the cycle.

Periodic Systems are of utmost relevance in control applications

Periodic System: A system where the temporal behaviour is structured into a sequence of periods.

Periodicity is not mandatory, but often assumed as it leads to simpler algorithms and more stable and secure systems [14], pp. 19–4. Note that the difference between *cycle* and *period* is the constant duration of the period during the IoD.

3.2 Clocks

Time is measured with clocks. In the cyber-domain, digital clocks are used.

Clock: A (digital) clock is an autonomous system that consists of an oscillator and a register. Whenever the oscillator completes a period, an event is generated that increments the register.

Oscillators and digital clocks are closely related. When looking at an oscillator, the form of the wave over the full cycle is of importance. When looking at a clock, only the distance between the events that denote the completion of cycles of the oscillator is of importance.

Nominal Frequency: The desired frequency of an oscillator [24].

Frequency drift: A systematic undesired change in frequency of an oscillator over time [24].

Frequency drift is due to ageing plus changes in the environment and other factors external to the oscillator.

Frequency offset: The frequency difference between a frequency value and the reference frequency value [24].

Stability: The stability of a clock is a measure that denotes the constancy of the oscillator frequency during the IoD.

The state of the register of a clock is often called the *state of clock*. The state of a clock remains constant during a complete period of the oscillator.

Wander: The long-term phase variations of the significant instants of a timing signal from their ideal position on the time-line (where long-term implies here that these variation of frequency are less than 10 Hz). (see also jitter) [24].

Jitter: The short-term phase variations of the significant instants of a timing signal from their ideal position on the time-line (where long-term implies here that these variation of frequency are greater than or equal to 10 Hz). (see also wander) [24].

The term timing signal can refer to a signal of a clock or of any other periodic event. There exist other clocks, e.g., a *sun dial*, which is not digital in nature. The time resolution of every digital clock is limited by the duration of the period of the oscillator.

Tick: The event that increments the register is called the tick of the clock.

Granularity/Granule of a clock: The duration between two successive ticks of a clock is called the granularity of the clock or a granule of time.

The granularity of a clock can only be measured if another clock with a finer granularity is available. We introduce a *reference clock* as a *working hypothesis* for measuring the instant of occurrence of an event of interest (such as, e.g., a clock tick) and make the following three hypothetical assumptions: (i) the reference clock has such a small granularity, e.g., a *femto second* (10^{-15} s), that digitalization errors can be neglected as second order effects, (ii) the reference clock can observe every event of interest without any delay and (iii) the state of the reference clock is always in perfect agreement with TAI time.

Reference clock: A hypothetical clock of a granularity smaller than any duration of interest and whose state is in agreement with TAI.

Coordinated Clock: A clock synchronized within stated limits to a reference clock that is spatially separated [24].

Every *good (fault-free) free-running clock* has an individual granularity that can deviate from the specified *nominal granularity* by an amount that is contained in the specification document of the physical clock under investigation.

Drift: The drift of a physical clock is a quality measure describing the frequency ratio between the physical clock and the reference clock.

Since the drift of a good clock is a number close to 1, it is conducive to introduce a drift rate by

$$Drift\ Rate = |Drift - 1|$$

Typical clocks have a drift rate of 10^{-4} to 10^{-8}. There exists no perfect clock with a drift rate of 0. The drift rate of a good clock will always stay in the interval contained in the specification document of the clock. If the drift rate of a clock leaves this specified interval, we say that the clock has *failed*.

Timestamp (of an event): The timestamp of an event is the state of a selected clock at the instant of event occurrence.

Note that a timestamp is always associated with a selected clock. If we use the reference clock for time-stamping, we call the time-stamp *absolute*.

Absolute Timestamp: An absolute timestamp of an event is the timestamp of this event that is generated by the reference clock.

If events are occurring close to each other, closer than the granularity of a digital clock, then an existing temporal order of the events cannot be established on the basis of the timestamps of the events.

If two events are *timestamped* by two different clocks, the temporal order of the events can be established on the basis of their timestamps only if the two clocks are synchronized.

Clock Ensemble: A collection of clocks, not necessary in the same physical location, operated together in a coordinated way either for mutual control of their individual properties or to maximize the performance (time accuracy and frequency stability) and availability of a time-scale derived from the ensemble [24].

Clock synchronization establishes a *global notion of time* in a clock ensemble. A global notion of time is required in an SoS if the timestamps generated in one CS must be interpreted in another CS. *Global time* is an abstraction of physical time in a distributed computer system. It is approximated by a properly selected subset *of the ticks* of each synchronized local clock of an ensemble. A *selected tick* of a local *clock* is called a *tick* of the *global time*. For more information on the global notion of time, see [2], pp. 58–64.

Precision: The precision of an ensemble of synchronized clocks denotes the maximum offset of respective ticks of the global time of any two clocks of the ensemble over the IoD. The precision is expressed in the number of ticks of the reference clock.

The precision of an ensemble of clocks is determined by the quality of the oscillators, by the frequency of synchronization, by the type of synchronization algorithm and by the jitter of the synchronization messages. Once the precision of the ensemble has been established, the granularity of the global time follows by applying the *reasonableness condition*.

Reasonableness Condition: The reasonableness condition of clock synchronization states that the granularity of the global time must be larger than the precision of the ensemble of clocks.

We distinguish between two types of clock synchronization, internal clock synchronization and external clock synchronization.

Internal Clock Synchronization: The process of mutual synchronization of an ensemble of clocks in order to establish a global time with a bounded precision.

There are a number of different internal synchronization algorithms, both non-fault tolerant or fault-tolerant, published in the literature (see e.g., [13], and many others). These algorithms require the cooperation of all involved clocks.

External Clock Synchronization: The synchronization of a clock with an external time base such as GPS.

Primary Clock: A clock whose rate corresponds to the adopted definition of the second. The primary clock achieves its specified accuracy independently of calibration.

The term *master clock* is often used synonymously to the term *primary clock.* If the clocks of an ensemble are externally synchronized, they are also internally synchronized with a precision of $|2A|$, where A is the *accuracy*.

Accuracy: The accuracy of a clock denotes the maximum offset of a given clock from the external time reference during the IoD, measured by the reference clock.

The external time reference can be a primary clock or the GPS time.

3.3 Time in an SoS

In a recent report from the GAO to the US Congress [15] it is noted that *a global notion of time is required in nearly all infrastructure SoSs,* such as telecommunication, transportation, energy, etc. In an SoS, *external clock synchronization* is the preferred alternative to establish a global time, since the scope of an SoS is often ill defined and it is not possible to identify *a priori* all CSs that must be involved in the (internal) clock synchronization. A CS that does not share the global time established by a subset of the CSs cannot interpret the timestamps that are produced by this subset.

The preferred means of clock synchronization in an SoS is the external synchronization of the local clocks of the CSs with the standardized time signal distributed worldwide by satellite navigation systems, such as GPS, Galileo or GLONASS. The GPS system, consisting at least of 24 active satellites transmit periodic time signals worldwide that are derived from satellite-local atomic clocks and seldom differ from each other by more than 20 ns [14]. A GPS receiver decodes the signals and calculates, based on the offset among the signals, the position and time at the location of the GPS receiver. The *accuracy* of the GPS time is better than 100 ns. The periodic time signal, generated by a GPS receiver, can be used to discipline a quartz oscillator.

GPSDO (Global Positioning System Disciplined Oscillator): The GPSDO synchronizes its time signals with the information received from a GPS receiver.

With a well-synchronized GPSDO a drift rate in the order 10^{-10} can be achieved.

Holdover: The duration during which the local clock can maintain the required precision of the time without any input from the GPS.

According to [16], p. 62 a good GPSDO has deviated from GPS time by less than 100 μsec during the loss of GPS input of one week. As long as the GPS is operating and its input is available, a GPSDO can provide an accuracy of the global time of better than 100 nsec. If there is the requirement that, the free running global time must not deviate by more than 1 μsec, a holdover of up to one hour is achievable using a good GPSDO.

The measurement of the position of an event on the timeline or of the duration between two events by a *digital global time* must take account of two types of unavoidable errors, the *synchronization error* caused by the finite precision of the global time and the *digitalization error* caused by the discrete time base. If the *reasonableness condition* is respected, the sum of these errors will be less than $2g$, where g is the granularity of the global time. It follows that the true duration between two events d_{true} lies in the following interval around the observed value d_{obs}.

$$(d_{obs} - 2g) < d_{true} < (d_{obs} + 2g)$$

The duration between events that are temporally ordered can be smaller than the granularity of a single clock. This situation is even worse if two different globally synchronized clocks observe the two different events. It is therefore impossible to establish the true temporal order of events in case the events are closer together than $2g$. This impossibility result can give rise to *inconsistencies* about the perceived temporal order of two events in distributed system.

These inconsistencies can be avoided, if a minimum distance between events is maintained, such that the temporal order of the events, derived from the timestamps of the events that have been generated by different clocks of a system with properly synchronized clocks is always the same.

Sparse Time: A time-base in a distributed computer system where the physical time is partitioned into an infinite sequence of active and passive intervals.

The active intervals can be enumerated by the sequence of natural numbers and this number can be assigned as the timestamp of an event occurring in an active interval. In order to establish consistency all events that occur in the same active interval of a sparse time are considered to have occurred simultaneously. This procedure establishes *consistency* at the price of *faithfulness*, since the temporal order of events that are closer together than the distance between sparse events is lost.

Sparse Events: Events that occur in the active interval of the sparse time.

Events that are in the SoC of a computer system with access to a global time, e.g., the start of sending a message, can be delayed until the next active interval and thus can be forced to be sparse events.

Non-Sparse Events: Events that occur in the passive interval of the sparse time.

Events that are outside the SoC of the computer system and are in the SoC of the environment cannot be forced to occur in the active intervals of the sparse time base, and can therefore be *non-sparse events*.

If all observers of a non-sparse event agree, by the execution of an agreement protocol, that the observed non-sparse event should be moved to the same nearest active interval of the sparse time base, then the consistency of these events can be established at the price of a further reduced faithfulness.

Time-aware SoS: A SoS is time-aware if its Constituent Systems (CSs) can use a global timebase in order to timely conduct output actions and consistently – within the whole SoS – establish the temporal order of observed events.

4 Data and State

Systems-of-Systems (SoSs) come about by the transfer of *information* of one Constituent System (CS) to another CS. But what is *information*? How is *information* related to *data*? After a thorough investigation of the literature about the fundamental concepts *data* and *information* it is concluded, that these terms are not well-defined in the domain of information science-see also the paper by C. Zins who asked a number of computer scientists about their meaning associated with the terms *data-information-knowledge* and published the divergent views reported to him in [16]. In this Section we will elaborate on the concepts of *data* and *information* along the lines of reasoning expressed in [17].

4.1 Data and Information

Let us start by defining the *fundamental* concepts of *data* and *information* [17]:

Data: A data item is an artefact, a pattern, created for a specified purpose.

In cyber space, data is represented by a *bit-pattern*. In order to arrive at the meaning of the bit pattern, i.e., the *information* expressed by the bit pattern, we need an *explanation* that tells us how to interpret the given bit pattern.

Information: A proposition about the state of or an action in the world.

A proposition can be about *factual circumstances* or *plans*, i.e., schemes for action in the future. In any case, information belongs to the category of *constructs*, i.e., non-physical entities in world three of Popper [8]. Sometimes the phrase *semantic content* is used as a synonym for information.

Explanation: The explanation of the data establishes the links between data and already existing concepts in the mind of a human receiver or the rules for handling the data by a machine.

Since only the combination of *data* and an associated *explanation* can convey information, we form the new concept of an *Itom* that we consider the smallest unit that can carry *information*.

Itom: An Itom (Information Atom) is a tuple consisting of data and the associated explanation of the data.

The concept of an *Itom* is related to the concept of an *infon* introduced by Floridi [18]. However, the properties we assign to an *Itom* are different from the properties Floridi assigns to an *infon*. The concept of an Itom does not make any assumptions about the truthfulness of the semantic content, the information, in the Itom. We thus can attribute factual information as true information (*correspondence theory of truth* [19]), misinformation (accidentally false) or disinformation (intentionally false), or just call it information if we do not know yet if it is true or false. It is often the case that only some time after data has been acquired it can be decided whether the information conveyed by the data is true or false (e.g., consider the case of a value error of a sensor).

When data is intended for a *human receiver* then the explanation must describe the data using concepts that are *familiar* to the intended human receiver. When data is intended for processing by a *machine*, the *explanation* consists of two parts, we call them *computer instructions* and *explanation of purpose* [17].

The *computer instructions* tell the computer system how the *data bit-string* is partitioned into syntactic chunks and how the syntactic chunks have to be stored, retrieved, and processed by the computer. This part of the *explanation* can thus be considered as a *machine program* for a (virtual) computer. Such a machine program is also represented by a bit-string. We call the data bit-string *object data* and the instruction bit-string that *explains* the object data, *meta data*.

A computer *Itom* thus contains digital *object data* and digital *meta data*. The recursion stops when the *meta data* is a sequence of well-defined machine instructions for the *destined* computer. In this case, the *design of the computer* serves as an *explanation for the meaning of the data.*

The second part of the explanation of an Itom, the *explanation of purpose,* is directed to humans who are involved in the design and operation of the computer system, since the *notion of purpose is alien to a computer system.* The *explanation of purpose* is part of the documentation of the cyber system and must be expressed in a form that is *understandable* to the human user/designer.

To facilitate the exchange of information among heterogeneous computer systems in the Internet, markup languages, such as the Extensible Markup Language XML [20], that help to explain the meaning of data have been developed. Since in XML the explanation is separated from the data, the explanation can be adopted to the context of use of the data. Markup languages provide a mechanism to support an explanation of data. In many situations the explanation of the data is taken implicitly from the context.

When data is moved from one CS to another CS of an SoS, the context may change, implying that an explanation that is context-dependent changes as well. Take the example of *temperature* expressed as a *number*. In one context (e.g., Europe) the number is interpreted as degrees Celsius, while in another context (e.g., the US) the number is interpreted as degrees Fahrenheit. If we do not change the number (the data) then the meaning of the Itom is changed when moving the data from one context to another context. The neglected context sensitivity of data has caused accidents in SoSs [21].

An observation of a dynamic entity is only complete if the instant, the *timestamp* of making the observation, is recorded as part of the explanation.

The timestamp of input data is determined by the termination instant of the sensing process.

No timestamp is needed in the explanation when the properties of the observed entity are static. In the context of our work, we are mostly interested in dynamic properties of systems.

4.2 State

Many systems store information about their interactions with the environment (since the start of a system with a clean memory) and use this information to influence their future behaviour.

State: The state of a system at a given instant is the totality of the information from the past that can have an influence on the future behaviour of a system.

A state is thus a valued data structure that characterizes the condition of a system at an instant. The concept of state is meaningless without a concept of time, since the distinction between past and future is only possible if the system is time-aware.

Stateless System: A system that does not contain state at a considered level of abstraction.

Statefull System: A system that contains state at a considered level of abstraction.

The variables that hold the stored state in a statefull system are called *state variables*.

State Variable: A variable that holds information about the state.

State Space: The state space of a system is formed by the totality of all possible values of the state variables during the IoD.

Instantaneous State Space: The state space of a system is formed by the totality of all possible values of the state variables at a given instant.

If we observe the progress of a system, we will recognize that the size of the instantaneous state space grows or shrinks as time passes. The instantaneous state space has a relative minimum at the start and end of an atomic action.

The size of the instantaneous state space is important if we want to restart a system after a failure (e.g., a corruption of the state by a transient fault). We have to repair the corrupted instantaneous state before we can reuse the system. Generally, the smaller the instantaneous state space at the instant of reintegration, the easier it is to repair and restart a system.

Most control systems are cyclic or even periodic systems.

Ground State: At a given level of abstraction, the ground state of a cyclic system is a state at an instant when the size of the instantaneous state space is at a minimum relative to the sizes of the instantaneous state spaces at all other instants of the cycle.

We call the instant during the cycle of a cyclic system where the size of the instantaneous state has a minimum the *ground state instant*.

Ground State Instant: The instant of the ground state in a cyclic system.

At the ground state instant all information of the past that is considered relevant for the future behaviour should be contained in a declared ground state data structure. At the ground state instant no *task* may be active and all communication channels are flushed. Ground state instants are ideal for reintegrating components that have failed.

Declared Ground State: A declared data structure that contains the relevant ground state of a given application at the ground state instant.

The *declared ground state* is essential for system recovery. The declared ground state contains only of those ground state variables that are considered *relevant* by the designer for the future operation of the system in the given application. Other ground state variables are considered non-relevant because they have only a minor influence on the future operation of the system. The decision of whether an identified state variable is relevant or not relevant depends on a deep understanding of the dynamics of an application.

Concise State: The state of a system is considered concise if the size of the declared ground state is at most in the same order of magnitude as the size of the system's largest input message.

Many control systems have a concise state. There are other systems, such as data base systems that do not have a concise state—the size of the state of a data-base system can be Gigabytes.

In contrast to state variables that hold information about the state at an instant an event variable holds information about a *change* at an instant.

Event Variable: A variable that holds information about some change of state at an instant.

5 Actions and Behaviour

We can observe the dynamics of a system that consists of discrete variables by an *event-based view* or by a *state-based view*.

In the *event-based view* we observe the state of relevant state variables at the beginning of the observation and then record all events (i.e. changes of the state variables) and the time of occurrence of the events in a trace. We can reconstruct the value of all state variables at any past instant of interest by the recorded trace. However, if the number of events that can happen is not bounded, the amount of data generated by the event-based view cannot be bounded.

In the *periodic state-based view* (called *sampling*), we observe the values of relevant state variables at selected *observation instants* (the sampling points) and record these values of the state variables in a trace. The duration between two observation instants puts a limit on the amount of data generated by the state-based view. However, the price for this limit is the loss of fidelity in the trace. Events that happen within a duration that is shorter than the duration between the equidistant observation instants may get lost.

Sampling: The observation of the value of relevant state variables at selected observation instants.

Most control systems use *sampling* to acquire information about the controlled object. The choice of the *duration between two observation instants*, called *the sampling interval,* is critical for acquiring a satisfying image of the controlled object. This issue is discussed extensively in the literature about control engineering [23].

5.1 Actions

In the following we introduce some concepts that allow us to describe the dynamics of a computer system.

Action: The execution of a program by a computer or a protocol by a communication system.

An action is started at a specified instant by a start signal and terminates at a specified instant with the production of an end signal.

Start signal: An event that causes the start of an action.

End signal: An event that is produced by the termination of an action.

Between the start signal and end signal an action is active.

Execution Time: The duration it takes to execute a specific action on a given computer.

The execution time depends on the performance of the available hardware and is also data dependent.

Worst Case Execution Time (WCET): The worst-case data independent execution time required to execute an action on a given computer.

There are two possible sources for a start signal of an action.

Time-triggered (TT) Action: An action where the start signal is derived from the progression of time.

An action can also be started by the completion of the previous action or by some other event (e.g., the push of a start button).

Event-triggered (ET) Action: An action where the start signal is derived from an event other than the progression of time.

We distinguish also between computational actions and communication actions.

Computational Action: An action that is characterized by the execution of a program by a machine.

Communication Action: An action that is characterized by the execution of a communication protocol by a communication system.

In our model of an action we assume that at the start event an action reads input data and state. At the end event an action produces output data and a new state. An action *reads* input data if the input data is still available after the action. An action *consumes* input data if the input data record is unavailable after the consumption by an action.

An action *writes* output data, if an old version of the output data is *overwritten* by the output data generated by the action. An action *produces* output data if a *new unit* of output data is generated by the action.

Input Action: An action that reads or consumes input data at an interface.

Output Action: An action that writes or produces output data at an interface.

We distinguish between actions that access the state of a system and those that do not.

Stateless Action: An action that produces output on the basis of input only and does not read, consume, write or produce state.

Statefull action: An action that reads, consumes, writes or produces state.

An action starts at the *start signal* and terminates by producing an *end signal*. In the interval <*start signal, end signal*> an action is *active*. While an action is active, the notion of state is undefined.

We can compose actions to form action sequences.

Action Sequence: A sequence of actions, where the end-signal of a preceding action acts as the start signal of a following action.

An *action sequence* is often called a *process.*

Activity Interval: The interval between the start signal and the end signal of an action or a sequence of related actions.

An action at a given level of abstraction, e.g., the execution of a program, can be decomposed into sub-actions. The decomposition ends when the internal behaviour of a sub-action is of no concern.

Atomic Action: An atomic action is an action that has the all-or-nothing property. It either completes and delivers the intended result or does not have any effect on its environment.

Atomic actions are related to the notion of a component introduced above: neither the internals of components (from the point of view of structure) nor the internals of an atomic action (from the point of view of behaviour) are of interest.

Irrevocable Action: An action that cannot be undone.

An irrevocable action has a lasting effect on the environment of a system. For example, consider an output action that triggers an airbag in a car.

Idempotent Action: An action is idempotent if the effect of executing it more than once has the same effect as of executing it only once.

For example, the action *move the door to 45°* is idempotent, while the action *move the door by five degrees* is not idempotent. Idempotent actions are of importance in the process of recovery after a failure.

We can combine computational action and communication actions to form a transaction.

Transaction: A related sequence of computational actions and communication actions.

Real-Time (RT) Transaction: A time-bounded transaction.

In a control system the duration of the RT-transaction that starts with the observation of the controlled object and terminates with the output of the result to an actuating device has an effect on the quality of control.

Transaction Activity Interval: The interval between the start signal and the end signal of a transaction.

5.2 Behaviour

The behaviour of a system-the observable traces of activity at the system interfaces-is of utmost interest to a user.

Function: A function is a mapping of input data to output data.

Behaviour: The timed sequence of the effects of input and output actions that can be observed at an interface of a system.

The effect of a *consuming input action* is the consumption of the input data record, while the effect of a reading input action does not change the input data and therefore has no observable effect. This is not true at the output side. Both, a *writing output action* and a *producing output action* have an observable effect.

Deterministic Behaviour: A system behaves deterministically if, given an initial state at a defined instant and a set of future timed inputs, the future states, the values and instants of all future outputs are entailed.

A system may exhibit an intended behaviour or it may demonstrate a behaviour that is unintended (e.g., erroneous behaviour).

Service: The intended behaviour of a system.

The service specification must specify the intended behaviour of a system. In non real-time systems, the service specification focuses on the data aspect of the behaviour. In real-time systems the *precise temporal specification* of the service is an integral part of the specification.

Capability: Ability to perform a service or function.

6 Communication

It is the basic objective of a communication system to transport a message from a sender to one or more receivers *within a given duration* and with a *high dependability*. By *high dependability* we mean that by the end of a specified time window the message should have arrived at the receivers with a high probability, the message is not corrupted, either by unintentional or intentional means, and that the security of the message (confidentiality, integrity, etc.) has not been compromised. In some environments e.g., the Internet of Things (IoT), there are other constraints on the message transport, such as, e.g., minimal energy consumption.

In an SoS the communication among the CSs by the exchange of messages is the core mechanism that realizes the integration of the CSs. It is imperative to elaborate on the concepts related to the message exchange with great care.

Since communication requires that diverse senders and receivers agree on the rules of the game, all involved partners must share these rules and their interpretation.

Communication Protocol: The set of rules that govern a communication action.

In the past fifty years, hundreds of different communication protocols that often only differ in minor aspects, have been developed. Many of them are still in use in legacy systems. This diversity of protocols hinders the realization of the economies of scale by the semiconductor industry.

We distinguish between two classes of protocols: basic transport protocols and higher-level protocols. The basic transport protocols are concerned with the transport of data from a sender to one or more receivers. The higher-level protocols build on these basic transport protocols to provide more sophisticated services.

6.1 Messages

Message: A data structure that is formed for the purpose of the timely exchange of information among computer systems.

We have introduced the word *timely* in this definition to highlight that a message combines concerns of the value domain and of the temporal domain in a single unit.

In the temporal domain, two important instants must be considered.

Send Instant: The instant when the first bit of a message leaves the sender.

Arrival Instant: The instant when the first bit of a message arrives at the receiver.

Receive Instant: The instant when the last bit of a message arrives at the receiver.

Transport Duration: The duration between the send instant and the receive instant.

Messages can be classified by the strictness of the temporal requirements.

In real-time communication systems strict deadlines that limit the transport duration must be met by the communication system.

From the point of view of the value domain, a message normally consists of three fields: a *header*, a *data field*, and a *trailer*. The header contains transport information that is relevant for the transport of the message by the communication system, such as the delivery address, priority information etc. The data field contains the payload of a message that, from the point of view of transport, is an *unstructured bit vector*. The trailer contains redundant information that allows the receiver to check whether the bit vector in the message has been corrupted during transport. Since a corrupted message is discarded, we can assume (as the fault model on a higher level) that the communication system delivers either a correct message or no message at all.

6.2 Basic Transport Service

A basic transport service transfers a message from a sender to one or more receivers. We limit our discussion to three different transport protocol classes that are representative for a number of important protocols in each class:

- Datagram
- PAR Message
- TT Message

Datagram: A best effort message transport service for the transmission of sporadic messages from a sender to one or many receivers.

A datagram is a very simple transport service. A datagram is forwarded along the best available route from a sender to one or a number of receivers. Every datagram is considered independent from any other datagram. It follows that a sequence of datagrams can be reordered by the communication system. If a datagram is corrupted or lost, no error will be indicated.

PAR-Message: A PAR-Message (Positive Acknowledgment or Retransmission) is an error controlled transport service for the transmission of sporadic messages from a sender to a single receiver.

In *the positive acknowledgment-or-retransmission* (PAR) protocol a sender waits for a given time until it has received a positive acknowledgement message from the receiver indicating that the previous message has arrived correctly. In case the timeout elapses before the acknowledgement message arrives at the sender, the original message is retransmitted. This procedure is repeated n-times (protocol specific) before a permanent failure of the communication is reported to the high-level sender. The jitter of the PAR protocol is substantial, since in most cases the first try will be successful, while in a few cases the message will arrive after n times the timeout value plus the worst-case message transport latency. Since the timeout value must be longer than two worst-case message transport latencies (one for the original message and one for the acknowledgment) the jitter of PAR is longer than $(2n)$ worst-case message-transport latencies ([2], p. 169). In addition to this basic PAR protocol one can find many protocol variants that refine this basic PAR protocol.

TT-Message: A TT-Message (Time-Triggered) is an error controlled transport service for the transmission of periodic messages from a sender to many receivers where the send instant is derived from the progression of the global time.

A time-triggered message is a periodic message that is transported from one sender to one or more receivers according to a pre-planned schedule. Since it is known *a priori* at the sender, the receiver and the communication system when a time-triggered message is expected to arrive, it is possible to avoid conflicts and realize a tight phase alignment between an incoming and an outgoing message in a communication system switch. The error detection of a TT-message is performed by the receiver on the basis of his a priori knowledge about the expected arrival time of a message. The error detection latency is determined by the precision of the global time.

Table 2 reports on the main characteristics of the transport services surveyed above. Although the basic datagram service does not provide temporal error detection, a-posteriori error detection of datagram messages can be achieved by putting the send timestamp in the message, given that synchronized clocks are available.

It is up to the application to decide which basic transport protocol is most appropriate to integrate the CSs into an SoS.

Table 2. Characteristics of transport services

Characteristic	Datagram	PAR-message	TT-message
Send Instants	sporadic	sporadic	periodic
Data/Control Flow	uni-directional	bi-directional	uni-directional
Flow Control	none	explicit	implicit
Message Handling	R/W or C/P	C/P	R/W
Transport Duration	a priori unknown	upper limit known	tight limit known
Jitter of the Message	unknown	large	small
Temporal Error Detection	none	at Sender	at Receiver
Example	UDP	TCP/IP	TT-Ethernet

6.3 High-Level Protocols

The transport protocols form the basis for the design of higher-level protocols, such as protocols for *file transmission* and *file sharing*, *device detection* and numerous other tasks. It is beyond the scope of this conceptual model to discuss the diversity of higher-level protocols. In the latter part of the AMADEOS project we will look at some of the higher level protocols that are of particular relevance in SoSs.

However, it must be considered that the temporal properties of the basic transport protocol determine to a significant extent the temporal properties of the high-level protocols.

6.4 Stigmergy

Constituent systems (CSs) that form the autonomous subsystems of Systems-of-Systems (SoSs) can exchange information items via two different types of channels: the conventional communication channels for the transport of messages and the *stigmergic channels* that transport information via the change and observation of states in the environment. The characteristics of the stigmergic channels, which often close the missing link in a control loop can have a decisive influence on the system-level behaviour of an SoS and the appearance of emergent phenomena [26].

Stigmergy: Stigmergy is a mechanism of indirect coordination between agents or actions. The principle is that the trace left in the environment by an action stimulates the performance of a next action, by the same or a different agent.

The concept of *stigmergy* has been first introduced in the field of biology to capture the indirect information flow among ants working together [27, 28]. Whenever an ant builds or follows a trail, it deposits a greater or lesser amount of pheromone on the trail, depending on whether it has successfully found a prey or not. Due to positive feedback, successful trails—i.e., trails that lead to an abundant supply of prey—end up with a high concentration of pheromone. The running speed of the ants on a trail is a non-linear function of the trail-pheromone concentration. Since the trail-pheromone evaporates—we call this process *environmental dynamics*—unused trails disappear autonomously as time progresses.

Environmental Dynamics: Autonomous environmental processes that cause a change of state variables in the physical environment.

The impact of environmental dynamics on the stigmergic information ensures that the captured itoms are well-aligned with the current state of the physical environment. No such alignment takes place if itoms are transported on cyber channels.

Stigmergic Information Flow: The information flow between a sending CS and a receiving CS where the sending CS initiates a state change in the environment and the receiving CS observes the new state of the environment.

If the output action of the sender and the input action of the receiver are closing a stigmergic link of a control loop, then the synchronization of the respective output and input actions and the transfer function of the physical object in the environment determine the delay of the stigmergic link and are thus of importance for the performance and stability of the control loop. The synchronization of the respective output and input actions requires the availability of a global time base of known precision.

A good example for a stigmergic information flow is the exchange of information among the drivers of cars on a busy intersection.

7 Interfaces

Central to the integration of systems are their interfaces, i.e., their points of interaction with each other and the environment over time. A point of interaction allows for an exchange of information among connected entities.

Interaction: An interaction is an exchange of information at connected interfaces.

The concept of a channel represents this exchange of information at connected interfaces.

Channel: A logical or physical link that transports information among systems at their connected interfaces.

A channel is implemented by a communication system (e.g., a computer network, or a physical transmission medium) which might affect the transported information, for example by introducing uncertainties in the value/time domains. In telecommunications a channel model describes all channel effects relevant to the transfer of information.

Interface Properties: The valued attributes associated with an interface.

Interface Layer: An abstraction level under which interface properties can be discussed.

Cyber Space: Cyber space is an abstraction of the Universe of Discourse (UoD) that consists only of information processing systems and cyber channels to realize message-based interactions.

Environmental Model: A model that describes the behavior of the environment that is relevant for the interfacing entities at a suitable level of abstraction.

Note that abstraction is always associated with a given specified purpose.
Interface properties can be characterized at different interface layers:

- **Cyber-Physical Layer:** At the cyber-physical layer information is represented as data items (e.g., a bit-pattern in cyberspace, or properties of things/energy in the physical world) that are transferred among interacting systems during the IoD.
- **Itom Layer:** In this layer we are concerned with the timely exchange of Itoms by unidirectional channels across CS interfaces.
- **Service Layer:** At the service layer, the interface exposes the system behavior structured as *capabilities*. In contrast to the informational layer, Itom channels are not individually described at the service layer, but only the interdependencies between the exchanged Itoms are specified.

Each system's point of interaction is an interface that (1) together with all other system interfaces establishes a well-defined boundary of the system, and (2) makes system services to other systems or the environment available. Consequently, any possibly complex internal structure that is responsible for the observable system behaviour can be reduced to the specification of the system interfaces [2].

Interface Specification: The interface specification defines at all appropriate interface layers the interface properties, i.e., what type of, how, and for what purpose information is exchanged at that interface.

7.1 System-of-Systems Interfaces

Interfaces within Constituent Systems (CSs) that are not exposed to other CSs or the CS's environment are internal.

Internal Interface: An interface among two or more subsystems of a Constituent System (CS).

External Interface: A Constituent System (CS) is embedded in the physical environment by its external interfaces.

We distinguish three types of external CS interfaces: Time-Synchronization Interface (TSI), Relied Upon Interfaces (RUIs) and *utility interfaces*. RUIs have been defined in Sect. 2.3.

Time-Synchronization Interface (TSI): The TSI enables external time-synchronization to establish a global timebase for time-aware CPSoSs.

Utility Interface: An interface of a CS that is used for the configuration, the control, or the observation of the behaviour of the CS.

The purposes of the utility interfaces are to (1) configure and update the system, (2) diagnose the system, and (3) let the system interact with its remaining local physical environment which is unrelated to the services of the SoS. In acknowledgement of these three purposes we introduce the utility interfaces: Configuration Interface (C-Interface), Diagnostic Interface (D-Interface), and Local I/O Interface (L-Interface).

Configuration and Update Interface (C-Interface): An interface of a CS that is used for the integration of the CS into an SoS and the reconfiguration of the CS's RUIs while integrated in a SoS.

The C-Interface is able to modify the interface specification of RUIs. If we can rely on a SoS where the CSs have access to a global time base, we can allow non-backward compatible updates (i.e., discontinuous evolution) and more importantly support time-controlled SoS evolution. A predefined validity instant which is part of the interface specification determines when all affected CSs need to use the updated RUI specification and abandon the old RUI specification. This validity instant should be chosen appropriately far in the future (e.g., in the order of the update/maintenance cycle of all impacted CSs).

Validity Instant: The instant up until an interface specification remains valid and a new, possibly changed interface specification becomes effective.

Service providers guarantee that the old interface specification remains active until the validity instant such that service consumers can rely on them up to the reconfiguration instant.

Diagnosis Interface (D-Interface): An interface that exposes the internals of a Constituent System (CS) for the purpose of diagnosis.

The D-Interface is an aid during CS development and the diagnosis of CS internal faults.

Monitoring CS: A CS of an SoS that monitors the information exchanges across the RUMIs of an SoS or the operation of selected CSs across the D-Interface.

There are interfaces among the components of a CS which are hidden behind the RUI of the CS and which are not visible from the outside of a CS, e.g., the interface between a physical sensor and the system that captures the raw data, performs data conditioning and presents the refined data at the RUMI.

Local I/O Interface (L-Interface): An interface that allows a Constituent System (CS) to interact with its surrounding physical reality or other CSs that is not accessible over any other external interface.

Some external interfaces are always connected with respect to the currently active operational mode of a correct system.

Connected Interface: An interface that is connected to at least one other interface by a channel.

A disconnected external interface might be a fault (e.g., a loose cable that was supposed to connect a joystick to a flight control system) which causes possibly catastrophic system failure.

In a SoS the CSs may connect their RUIs according to a RUI connecting strategy that searches for and connects to RUIs of other CSs.

RUI connecting strategy: Part of the interface specification of RUIs is the RUI connecting strategy which searches for desired, w.r.t. connections available, and compatible RUIs of other CSs and connects them until they either become undesirable, unavailable, or incompatible.

One important class of faults that might occur at connected interfaces is related to compatibility.

Property Mismatch: A disagreement among connected interfaces in one or more of their interface properties.

Connection System/Gateway Component/Wrapper: A new system with at least two interfaces that is introduced between interfaces of the connected component systems in order to resolve property mismatches among these systems (which will typically be legacy systems), to coordinate multicast communication, and/or to introduce emerging services.

8 Evolution and Dynamicity

Large scale Systems-of-Systems (SoSs) tend to be designed for a long period of usage (10 years+). Over time, the demands and the constraints put on the system will usually change, as will the environment in which the system is to operate. The AMADEOS project studies the design of systems of systems that are not just robust to dynamicity (short term change), but to long term changes as well. This Section addresses a number of terms related to the evolution of SoSs.

Evolution: Process of gradual and progressive change or development, resulting from changes in its environment (primary) or in itself (secondary).

Although the term evolution in other contexts does not have a positive or negative direction, in the SoSs context, evolution refers to maintaining and optimizing the system - a positive direction, therefore.

Managed evolution: Evolution that is guided and supported to achieve a certain goal.

For SoSs, evolution is needed to cope with changes. Managed evolution refers to the evolution guidance. The goal can be anything like performance, efficiency, etc. The following two definitions further detail managed evolution for SoSs:

Managed SoS evolution: Process of modifying the SoS to keep it relevant in face of an ever-changing environment.

This is Primary evolution; examples of environmental changes include new available technology, new business cases/strategies, new business processes, changing user needs, new legal requirements, compliance rules and safety regulations, changing political issues, new standards, etc.

Unmanaged SoS evolution: Ongoing modification of the SoS that occurs as a result of ongoing changes in (some of) its CSs.

This is Secondary evolution; examples of such internal changes include changing circumstances, ongoing optimization, etc. This type of evolution may lead to unintended emergent behaviour, e.g., due to some kind of "mismatch" between Constituent Systems (CSs) (see Sect. 2.3).

Local Evolution: Local evolution only affects the internals of a Constituent System (CS) which still provides its service according to the same and unmodified Relied Upon Interface (RUI) specification.

Global Evolution: Global evolution affects the SoS service and thus how CSs interact. Consequently, global evolution is realized by changes to the Relied Upon Interface (RUI) specifications.

Evolutionary Performance: A quality metric that quantifies the business value and the agility of a system.

Evolutionary Step: An evolutionary change of limited scope.

Minor Evolutionary Step: An evolutionary step that does not affect the Relied Upon Interface (RUI) Itom Specification (I-Spec) and consequently has no effects on SoS dynamicity or SoS emergence.

Major Evolutionary Step: An evolutionary step that affects the Relied Upon Interface (RUI) Itom specification and might need to be considered in the management of SoS dynamicity and SoS emergence.

Managed SoS evolution is due to changes in the environment. The goal of managed evolution in AMADEOS is maximizing business value, while maintaining high SoS agility:

Business value: Overarching concept to denote the performance, impact, usefulness, etc. of the functioning of the SoS.

Agility (of a system): Quality metric that represents the ability of a system to efficiently implement evolutionary changes.

Quantizing business value is difficult since it is a multi-criteria optimization problem. Aiming for Pareto optimality, various aspects (measured by utility functions) are weighted on a case-by-case basis.

System performance is a key term in the concept of business value:

System performance: The combination of system effectiveness and system efficiency.

System effectiveness: The system's behaviour as compared to the desired behaviour.

System efficiency: The amount of resources the system needs to act in its environment.

For the last definition, it is important to understand what system resources are in the area of SoS:

System resources: Renewable or consumable goods used to achieve a certain goal. E.g., a CPU, CPU-time, electricity.

Dynamicity of a system: The capability of a system to react promptly to changes in the environment.

Linked to dynamicity and to the control strategy shifting from a central to an autonomous paradigm, is the concept of *reconfigurability*.

Reconfigurability: The capability of a system to adapt its internal structure in order to mitigate internal failures or to improve the service quality.

We conclude the discussion presenting fundamentals on governance, because governance-related facts may have impact on SoS evolution.

Authority: The relationship in which one party has the right to demand changes in the behaviour or configuration of another party, which is obliged to conform to these demands.

(Collaborative) SoS Authority: An organizational entity that has societal, legal, and/or business responsibilities to keep a collaborative SoS relevant to its stakeholders. To this end it has authority over RUI specifications and how changes to them are rolled out.

For the purpose of rolling out changes to RUI specifications, the SoS authority needs the capabilities to measure the state of the implemented changed RUI specifications, and to give incentives to motivate CSs to implement the RUI specification changes.

Incentive: Some motivation (e.g., reward, punishment) that induces action.

9 System Design and Tools

SoSs can have a very complex architecture. They are constituted by several CSs which interact with each other. Due to their complexity, well defined design methodologies should be used, in order to avoid that some SoS requirement is not fulfilled and to ease the maintainability of the SoS.

9.1 Architecture

The architecture of a system can have some variants or even can vary during its operation. We recognize this adaptability of a system architecture by the following three concepts.

Evolvable architecture: An architecture that is adaptable and then is able to incorporate known and unknown changes in the environment or in itself.

Flexible architecture: Architecture that can be easily adapted to a variety of future possible developments.

Robust architecture: Architecture that performs sufficiently well under a variety of possible future developments.

The architecture then involves several components which interact with each other. The place where they interact is defined as interface.

During the development lifecycle of a system, we start from conceptual thoughts which are then translated into requirements, which are then mapped into an architecture. The process that brings designers to define a particular architecture of the system is called design.

Design: The process of defining an architecture, components, modules and interfaces of a system to satisfy specified requirement.

In the AMADEOS context, design is a verb, architecture is a noun. The people who perform the design are designers.

Designer: An entity that specifies the structural and behavioral properties of a design object.

There are several methodologies to design a system.

Hierarchical Design: A design methodology where the envisioned system is intended to form a holarchy or formal hierarchy.

Top Down Design: A hierarchical design methodology where the design starts at the top of the holarchy or formal hierarchy.

Bottom Up Design: A hierarchical design methodology where the design starts at the bottom of the holarchy or formal hierarchy.

Meet-in-the-Middle Design: A hierarchical design methodology where the top down design and the bottom up design are intermingled.

This methodology is useful to decrease the degree of the complexity of a system and to ease its maintainability. In particular, in the hierarchical design the system can be split in different subsystems that can be grouped in modules.

Module: A set of standardized parts or independent units that can be used to construct a more complex structure.

The modularity is the technique that combines these modules in order to build a more complex system.

Modularity: Engineering technique that builds larger systems by integrating modules.

In the context that an SoS shall deal with evolving environments, then also design methodologies focused on evolution shall be defined.

Design for evolution: Exploration of forward compatible system architectures, i.e. designing applications that can evolve with an ever-changing environment. Principles of evolvability include modularity, updateability and extensibility. Design for evolution aims to achieve robust and/or flexible architectures.

Examples for design for evolution are Internet applications that are forward compatible given changes in business processes and strategies as well as in technology (digital, genetic, information-based and wireless).

In the context of SoS, design for evolution can be reformulated in the following manner.

Design for evolution in the context of SoS: Design for evolution means that we understand the user environment and design a large SoS in such a way that expected changes can be accommodated without any global impact on the architecture. 'Expected' refers to the fact that changes will happen, it does not mean that these changes themselves are foreseeable.

In addition, during the system development lifecycle some activities to verify if the architecture developed during the design process is compliant and if it fulfills the requirements of the system are foreseen. Verification of design is an important activity especially in critical systems and several methodologies are defined to perform it. In particular there is a methodology which has in mind verification since the design phase.

Design for testability: The architectural and design decisions in order to enable to easily and effectively test our system.

Then we have two methodologies to perform design verification:

Design inspection: Examination of the design and determination of its conformity with specific requirements.

Design walkthrough: Quality practice where the design is validated through peer review.

In order to perform all these activities some useful tools can be used to help the designers and verifiers job.

10 Dependability and Security

We report the basic concepts related to dependability, security and multi-criticality. These are important properties for System-of-Systems (SoSs) since they impact availability/continuity of operations, reliability, maintainability, safety, data integrity, data privacy and confidentiality. Definitions for dependability and security concepts can be very subtle; slight changes in wording can change the entire meaning. Thus, we have chosen to refer to basic concepts and definitions that are widely used in the dependability and security community.

The reference taxonomy for the basic concepts of dependability applied to computer-based systems can be found in [3]. It is the result of a work originated in 1980, when a joint committee on "Fundamental Concepts and Terminology" was formed by the TC on Fault–Tolerant Computing of the IEEE CS1 and the IFIP WG 10.4 "Dependable Computing and Fault Tolerance" with the intent of merging the distinct but convergent paths of the dependability and security communities.

In addition to the work of Laprie [3, 4], we also refer to definitions from the *CNSS Instruction No. 4009: National Information Assurance (IA) Glossary* [1]. The CNSSI 4009 was created through a working group with the objective to create a standard glossary of security terms to be used across the U.S. Government, and this glossary is periodically updated with new terms. We also cite security terms as defined by Ross Anderson's "Security Engineering: A Guide to Building Dependable Distributed Systems" [5] and Bruce Schneier's "Applied Cryptography" [6].

10.1 Threats: Faults, Errors, and Failures

The threats that may affect a system during its entire life from a dependability viewpoint are failures, errors and faults. Failures, errors and faults are defined in the following, together with other related concepts.

Failure: The actual system behaviour deviation from the intended system behaviour.

Error: Part of the system state that deviated from the intended system state and could lead to system failure.

It is important to note that many errors do not reach the system's external interfaces, hence do not necessarily cause system failure.

Fault: The adjudged or hypothesized cause of an error; a fault is active when it causes an error, otherwise it is dormant.

The prior presence of vulnerability, i.e., a fault that enables the existence of an error to possibly influence the system behaviour, is necessary such that a system fails.

The creation and manifestation mechanism of faults, errors, and failures is called "chain of threats". The chain of threats summarizes the causality relationship between faults, errors and failures. A fault activates (fault activation) in component A and generates an error; this error is successively transformed into other errors (error propagation) within the component (internal propagation) because of the computation process. When some error reaches the service interface of component A, it generates a failure, so that the service delivered by A to component B becomes incorrect. The ensuing service failure of component A appears as an external fault to component B.

10.2 Dependability, Attributes, and Attaining Dependability

Dependability (original definition): The ability to deliver service that can justifiably be trusted.

The above definition stresses the need for justification of "trust", so an alternate definition is given:

Dependability (new definition): The ability to avoid failures that are more frequent and more severe than is acceptable.

This last definition has a twofold role, because in addition to the definition itself it also provides the criterion for deciding whether the system is dependable or not. Dependability is an integrating concept that encompasses the following dependability attributes:

Availability: Readiness for service.

Reliability: Continuity of service.

Maintainability: The ability to undergo modifications and repairs.

Safety: The absence of catastrophic consequences on the user(s) and on the environment.

Integrity: The absence of improper system state alterations.

A specialized secondary attribute of dependability is robustness.

Robustness: Dependability with respect to external faults (including malicious external actions).

The means to attain dependability (and security) are grouped into four major dependability categories:

Fault prevention: The means to prevent the occurrence or introduction of faults.

Fault prevention is part of general engineering and aims to prevent the introduction of faults during the development phase of the system, e.g. improving the development processes.

Fault tolerance: The means to avoid service failures in the presence of faults.

Fault tolerance aims to avoid the occurrence of failures by performing error detection (identification of the presence of errors) and system recovery (it transform a system state containing one or more errors into a state without detected errors and without faults that can be activated again) over time.

Fault removal: The means to reduce the number and severity of faults.

Fault removal can be performed both during the development phase, by performing verification, diagnosis and correction, and during the operational life, by performing corrective and preventive maintenance actions.

Fault forecasting: The means to estimate the present number, the future incidence, and the likely consequences of faults.

Fault forecasting is conducted by performing an evaluation of system behaviour with respect to fault occurrence of activation, using either qualitative evaluations (identifying, classifying and ranking the failure modes) or quantitative ones (evaluating in terms of probabilities the extent to which some of the attributes are satisfied).

The relationship among the above mentioned means are the following: fault prevention and fault tolerance aim to provide the ability to deliver a service that can be trusted, while fault removal and fault forecasting aim to reach confidence in that ability by justifying that the functional and the dependability and security specifications are adequate and that the system is likely to meet them.

10.3 Security

Continuing with the concepts defined by Laprie [4], when addressing security an additional attribute needs to be considered: confidentiality.

Confidentiality: The absence of unauthorized disclosure of information.

Based on the above definitions, security is defined as follows:

Security: The composition of confidentiality, integrity, and availability; security requires in effect the concurrent existence of availability for authorized actions only, confidentiality, and integrity (with "improper" meaning "unauthorized").

We also consider the security definitions proposed by the CNSSI 4009 [1], which discusses security in terms of risk management. In this glossary, security is a condition that results from the establishment and maintenance of protective measures that enable an enterprise to perform its mission or critical functions despite risks posed by threats to its use of information systems.

Threat: Any circumstance or event with the potential to adversely impact organizational operations (including mission, functions, image, or reputation), organizational assets, individuals, or other organizations through a system via unauthorized access, destruction, disclosure, modification of information, and/or denial of service.

A threat can be summarised as a failure, error or fault.

Vulnerability: Weakness in a system, system security procedures, internal controls, or implementation that could be exploited by a threat.

Vulnerabilities can be summarised as internal faults that enable an external activation to harm the system [4].

Risk is defined in terms of the impact and likelihood of a particular threat.

Risk: A measure of the extent to which an organization is threatened by a potential circumstance or event, and typically a function of (1) the adverse impacts that would arise if the circumstance or event occurs; and (2) the likelihood of occurrence.

Encryption using cryptography can be used to ensure confidentiality. The following definitions from Bruce Schneier's "Applied Cryptography" [6], have been adapted to incorporate our definitions of *data* and *information*.

Encryption: The process of disguising data in such a way as to hide the information it contains.

Cryptography: The art and science of keeping data secure.

Data that has not been encrypted is referred to as plaintext or cleartext. Data that has been encrypted is called ciphertext.

Decryption: The process of turning ciphertext back into plaintext.

Encryption systems generally fall into two categories, asymmetric and symmetric, that are differentiated by the types of keys they use.

Key: A numerical value used to control cryptographic operations, such as decryption and encryption.

Symmetric Cryptography: Cryptography using the same key for both encryption and decryption.

Public Key Cryptography (asymmetric cryptography): Cryptography that uses a public-private key pair for encryption and decryption.

11 Conclusions

The need of understanding and explaining the SoS in a clear way is a relevant requirement in order to reduce the cognitive complexity in SoS engineering. The scope of the set of concepts presented in this Chapter is to provide a common basis for the understanding and the description of Systems of Systems; we hope that such SoS concepts contribute towards such needed clarification.

Finally, we conclude mentioning that *SoS engineering* is closely related to many other IT domains by looking at the *glue* that is needed to integrate the diverse systems, for example *Cyber-Physical Systems (CPSs), embedded systems, Internet of Things (IoT), Big Data.* The concepts introduced in this Chapter and in the AMADEOS project to describe the properties of an SoS can thus form a foundation of a conceptual model in these other domains as well.

References

1. CNSS Instruction No. 4009: National Information Assurance (IA) Glossary, April 26, 2010. Retrieved from Committee on National Security Systems. http://www.ncix.gov/publications/policy/docs/CNSSI_4009.pdf
2. Kopetz, H.: Real-Time Systems: Design Principles for Distributed Embedded Applications, 2nd edn. Springer, NewYork (2011)
3. Laprie J.: Resilience for the scalability of dependability. In: Proceedings ISNCA 2005, pp. 5–6 (2005)
4. Avizienis, A., Laprie, J.-C., Randell, B., Landwehr, C.: Basic concepts and taxonomy of dependable and secure computing. IEEE Trans. Dependable Secure Comput. **1**(1), 11–33 (2004)
5. Anderson, R.: Security Engineering, 2nd edn. Wiley, New York (2008)
6. Schneier, B.: Applied Cryptography, 2nd edn. Wiley, New York (1996)
7. Jones, C., et al.: Final version of the DSoS conceptual model. DSoS Project (IST-1999-11585) (2002)
8. Popper, K.: Three Worlds: The Tanner Lecture on Human Values. University of Michigan (1978)

9. Simon, H.: The Science of the Artificial. MIT Press, Cambridge (1969)
10. Jamshidi, M.: Systems of Systems Engineering—Innovations for the 21st Century. Wiley, Cambridge (2009)
11. Dahman, J.S., Baldwin, K.J.: Understanding the current state of US defense systems of systems and the implications for systems engineering. In: Proceedings of 2nd Annual IEEE Systems Conference, Montreal. IEEE Press (2008)
12. Withrow, G.J.: The Natural Philosophy of Time. Oxford Science Publications, New York (1990)
13. Kopetz, H., Ochsenreiter, W.: Clock synchronization in distributed real-time systems. IEEE Trans. Comput. **36**(8), 933–940 (1987)
14. Lombardi, M.A.: The use of GPS disciplines oscillators as primary frequency standards for calibration and metrology laboratories. Measure J. Measur. Sci. **3**, 56–65 (2008)
15. US Government Accountability Office: GPS Disruptions: Efforts to Assess Risk to Critical Infrastructure and Coordinate Agency Actions Should be Enhanced. Washington, GAO-14-15 (2013)
16. Zins, C.: Conceptual approaches for defining data, information and knowledge. J. Am. Soc. Inf. Sci. Technol. **58**(4), 479–493 (2007)
17. Kopetz, H.A.: Conceptual model for the information transfer in systems of systems. In: Proceedings of ISORC 2014, Reno, Nevada. IEEE Press (2014)
18. Floridi, L.: Is semantic information meaningful data? Philos. Phenomenological Res. **60**(2), 351–370 (2005)
19. Glanzberg, M.: Truth: Stanford Encyclopedia on Philosophy (2013)
20. World Wide Web Consortium: Extensible Markup Language. http://www.w3.org/XML/. Accesssed 18 April 2013
21. Aviation Safety Network: Accident Description. http://aviationsafety.net/database/record.php?id=19920120-0 Accessed 10 June 2013
22. Maier, M.W.: Architecting principles for systems-of-systems. Syst. Eng. **1**(4), 267–284 (1998)
23. Zak, S.H.: Systems and Control. Oxford University Press, New York (2003)
24. International Telecommunication Union (ITU): Glossary and Definition of Time and Frequency Terms. Recommendation ITU-R TF686-3, December 2013
25. Frei, R., Di Marzo Serugendo, G.: Concepts in complexity engineering. Int. J. Bio-Inspired Comput. **3**(2), 123–139 (2011)
26. Kopetz, H.: Direct versus stigmeric information flow in systems-of-systems. TU Wien, November 2014 (not yet published)
27. Camazine, S., et al.: Self-Organization in Biological Systems. Princeton University Press, Princeton (2001)
28. Grasse, P.P.: La reconstruction du nid et les coordinations interindividuelles chez Bellicositermes natalensis et Cubitermes sp. La theorie de la stigmergie. Insectes Soc. **6**, 41–83 (1959)
29. Kopetz, H.: Real-Time Systems, 2nd edn. Springer, New York (2011)
30. Mogul, J.C.: Emergent (mis)behavior vs complex software systems. ACM SIGOPS Oper. Syst. Rev. **40**(4), 293–304 (2006)
31. Black, J., Koopman, P.: System safety as an emergent property in composite systems. In: 2009 IEEE/IFIP International Conference on Dependable Systems & Networks. IEEE (2009)
32. Mori, M., Ceccarelli, A., Zoppi, T., Bondavalli, A.: On the impact of emergent properties on SoS security. In: 7th International Conference on System of Systems Engineering (SoSE) (2016)
33. AMADEOS Consortium: D2.3 – AMADEOS Conceptual Model Revised (2016). http://amadeos-project.eu/documents/public-deliverables/

34. Gligor, V.: Security of emergent properties in ad-hoc networks (transcript of discussion). In: Christianson, B., Crispo, B., Malcolm, J.A., Roe, M. (eds.) Security Protocols 2004. LNCS, vol. 3957, pp. 256–266. Springer, Heidelberg (2006). doi:10.1007/11861386_30
35. DANSE Consortium: DANSE Methodology V2-D_4.3. https://www.danse-ip.eu
36. COMPASS, Guidelines for Architectural Modelling of SoS. Technical Note Number: D21.5a Version: 1.0, September 2014. http://www.compass-research.eu

Interfaces in Evolving Cyber-Physical Systems-of-Systems

Bernhard Frömel(✉) and Hermann Kopetz

Institute of Computer Engineering,
Vienna University of Technology, Vienna, Austria
froemel@vmars.tuwien.ac.at, h.kopetz@gmail.com

1 Introduction

In the past twenty years the view on how we engineer, operate and evolve independently owned and managed *Cyber-Physical Systems (CPSs)* in order to realize and optimize complex economical processes has started to change. Advances in telecommunications and automation accompanied by standardization efforts resulted in sophisticated cross-domain information and communication technologies (e.g., the Internet of Things (IoT) [2, 9], elastic processing and storage clouds, Web Services) that allow for the integration of more and more existing and previously technologically isolated CPSs. These *legacy systems* became cooperating *Constituent Systems (CSs)* of evolving Cyber-Physical Systems-of-Systems (CPSoSs) and – by their physical and cyber interaction – give rise to new emergent services that cannot be realized by any single or small number of CSs alone.

One prototypical example of a CPSoS is a smart grid [14] where the interacting CPSs (producers, consumers, and prosumers where, for example, electricity consuming households are equipped with electricity producing photovoltaic power plants) cooperate to optimize energy distribution with respect to stability, dependability, and costs. A smart grid handles high *dynamicity* as it constantly reconfigures in order to react to changed energy production and demand conditions. Further they need to support *evolution* during runtime as the *service* of the smart grid is adapted or extended towards new *requirements* or technological advances. Finally, smart grids represent critical infrastructure that may in the event of failure cost human lives or cause high economical costs. Hence a smart grid needs to fulfill high expectations concerning its *dependability*, including *security* and *safety*.

Central to the integration of CPSs as CSs of evolving CPSoSs are their *interfaces*, i.e., their points of interaction with each other (direct interaction) and with their common *environment* (indirect interaction) over *time*. The identification, proper specification, standardization, and managed modification of these interfaces are of paramount importance in order to tackle CPSoS key challenges related to emergence, dynamicity, evolution and dependability. Specifically, time-sensitive physical interactions and the role of delays in *emergence* impose the requirement of properly taking

This work has been partially supported by the FP7-610535-AMADEOS project.

A. Bondavalli et al. (Eds.): Cyber-Physical Systems of Systems, LNCS 10099, pp. 40–72, 2016.
DOI: 10.1007/978-3-319-47590-5_2

time for all kinds of interactions in CPSoSs into account. To this end this work assumes the availability of a *sparse global timebase* [25, 26] that can be used by all involved CSs to temporally coordinate *interactions* at their interfaces. We call an SoS where its CSs have access to such a global timebase a *time-aware SoS*.

The objective of this chapter is twofold: First, we conceptualize time-sensitive interactions in CPSoSs at appropriate *interface layers* and propose a CS interface design that simplifies engineering, operating and evolving such CPSoSs. Second, we discuss evolution of CPSoSs and how to manage it by applying our proposed interface design.

The following section gives a brief overview of related work. Section 3 conceptualizes interfaces in CPSoSs and introduces the *Relied Upon Interface (RUI)* of a CS which is an interface the operational service of the overall CPSoS relies upon. Section 4 discusses the design of RUIs. Section 5 suggests how evolution can be managed at the RUI. Section 6 concludes this chapter.

2 Related Work

In the domain of safety-critical real-time systems several simplification principles concerning the integration of models of distributed computer systems with models of physical processes have been suggested [25]: abstraction, separation of concerns, causality by determinism, temporal segmentation, independence of *entities*, observability, and consistent time. In accordance with these simplification strategies the design of linking interfaces [24] which realize cyber-interactions among nodes of a distributed real-time system, and the design of *sensor/actuator* interfaces [10] that interact with the physical process of a Cyber-Physical System (CPS) has been proposed.

Maier outlines in [30] fundamental differences in developing monolithic systems compared to a System-of-Systems (SoS) where for example the single Constituent Systems (CSs) are operationally and managerially independent, the SoS has an evolutionary nature, and there are emergent behaviors. Maier postulates that SoS architecting might rely entirely on interface design and the specification of communication standards at multiple abstraction levels.

The World Wide Web (WWW) running on top of the Internet is often considered as one of the first examples of a human engineered SoS, also satisfying Maier's definition of SoSs. Fielding [15] suggests the concept of an *architectural style*, i.e., in his thesis the *"named, coordinated set of architectural constraints"*, and uses it to obtain an appropriate architectural design for networked software components. Fielding further introduces the Representational State Transfer (REST) architectural style which he applied in the definition of the Hypertext Transfer Protocol (HTTP) and Uniform Resource Identifier (URI) specifications. Together they describe the generic interface which is in its essential form still used in all interactions of the WWW.

Web Services (WSs) [3, 8] are a prominent web-inspired implementation of the Service-oriented-Architecture (SoA). They are based on machine-interpretable interface descriptions (Web Service Description Language (WSDL), and other WS-* specifications) and give support for platform-independent machine-to-machine interaction over large-scale networks like the Internet. Highly distributed applications can be integrated by means of Web Service (WS) that are provided and owned by possibly many different entities. However, WSs are subject to the limitations of underlying technologies and the expressiveness of the used description languages. For example, the end-to-end *delay* of *messages* is in the scope of underlying technologies. Currently, temporal and semantic interoperability is not part of the *interface specification*, hence – while an agent might syntactically be able to interact with a WS – there might be a temporal and/or semantic mismatch leading to undesired effects.

OPC Unified Architecture (OPC UA) [29] is a SoA machine-to-machine communication stack intended to tackle challenges related to semantic interoperability. The OPC UA specifications are maintained by the Open Platform Communications (OPC) Foundation and have been in part standardized in IEC 62541. The extensible information model allows the description of arbitrary object structures where *object data* and its *meta data* is managed.

Caffall and Michael propose in [7] the concept of service-oriented contract interfaces for an architectural framework for SoSs. Contract interfaces are based on the principles of Design-by-Contract (DbC) and demand an explicit definition of interfaces among CSs that formalize assumptions about the services provided by a CS to achieve goals of the SoS.

3 Interfaces in Cyber-Physical Systems-of-Systems

This section introduces interfaces in the architectural context of time-aware Cyber-Physical Systems-of-Systems (CPSoSs), discusses interface abstraction layers, and presents different interface classes of Constituent Systems (CSs) that are part of a CPSoS.

3.1 Architectural Elements of Cyber-Physical Systems-of-Systems

A CPSoS is a System-of-Systems (SoS) whose interacting Constituent Systems (CSs) are Cyber-Physical Systems (CPSs). The architecture of a CPSoS defines the boundaries of its elements (major building blocks), the relationships among them, and the relationship between elements and their environment at abstraction levels that are useful for the discussion of CPSoS *attributes* of interest. There are many important attributes (e.g., business, societal, legal) to investigate in CPSoSs, but this chapter focuses on behavioral attributes, i.e., on interaction relations among architectural elements of CPSoSs. The architectural elements in CPSoSs are: the *Itom* [27] as the unit of interaction, the CS as a computing component that interacts physically and digitally with its environment, and the environment of CSs that enables the interactions among CSs. An Itom is an information atom comprising of data and explanation in an atomic unit. The environment of a CS consists of all entities that have the capability to interact with the CS.

CSs process Itoms which they exchange with their environment at their interfaces. There are two kinds of interactions a CS can have with its environment: message-based interactions with *cyber space* and physical or *stigmergic* interactions [28]. In order to model stigmergic interactions it is useful to define the concept of an *entourage* of a CS, i.e. all entities that are part of a CS, but are external of its computer system. It consists of humans and *things* and may change over time. The entourage of two or more CPSs may – not necessarily simultaneously – overlap during the *Interval of Discourse (IoD)*. Overlapping entourages provide a common environment which enables stigmergic interactions among involved CPSs.

Itom
An Itom [27] is the basic *information* item exchanged in interactions of CSs. It is defined as *"a timed proposition about some state or behavior in the world"* in some representation (data) together with the explanation of this representation. For processing in computer systems the representation and explanation can be both coded as bit strings, i.e., digital data. The representation part is called object data, while the explanation of the object data is called meta data. Depending on the Itom's purpose, the meta data might be based on an ontology (e.g., a conceptual model of Newtonian physics), or can be a machine program for a (virtual) computer system which explains the object data (cf. code mobility [17]). Interacting CSs may adhere to different *contexts*, i.e., often have implicit assumptions about the actual meaning and temporal *properties* of exchanged object data. Conceptually, Itoms solve this context dependency problem by tying information representation and explanation together. Hence object data is interpreted consistently across all CSs that need to access and process the information contained in the Itom.

The definition of Itoms is recursive and allows that an Itom contains other Itoms. Take for example the object- and meta data of two or more Itoms and regard them as object data of a higher level Itom. The explanation of this higher-level Itom describes how to extract the contained Itoms. Consequently, Itoms can be – in principle – arbitrarily complex constructs that might describe (virtual) computer machines and thus (virtual) CSs.

To avoid misinterpretation of interactions the explicit definition of Itoms as the basic unit of interaction is essential in modeling CPSoSs. Explicitly defined Itoms also enable automatic Itom *transformation systems* which are able to map object data in compliance to one explanation into object data conforming to another explanation and vice-versa, provided the two meta data explanations can be put formally into relation, i.e., a bijective function exists that maps one explanation (described by meta data) to the other.

Constituent System
A Constituent System (CS) is a Cyber-Physical System (CPS) which consists of a computer system (cyber part), optionally an influenced and/or observed object (physical part) and possibly humans. The object, i.e., a thing obeying to physical laws in the dense physical time of our reality, can be observed by sensors and/or influenced by actuators of the cyber part (for example for the purpose to control it). The *behavior* of the computer system and its ability to conduct *actions* in the physical environment adheres to a discrete progression of time. Consequently, a sufficient alignment of the dense physical time and

the discrete computer time is essential for an influence or *observation* of the physical part of a CPS that is precise enough for the application at hand.

The purpose of a CS regarding its integration in a *collaborative SoS* is the realization of a service that allows the CS to benefit from the emergent CPSoS service. The service of a CS is only provided at its interface thus an interface specification sufficiently defines the CS's interaction *capabilities* (all possible interactions) that the CS internals must deliver. In CPSoSs a precise temporal coordination of the individual CSs is required across the whole CPSoS. Hence, CSs have access to a sparse global timebase [25], and are able to *timestamp input actions* (messages, sensor observations) and timely generate *output actions* (messages, actuations) at the CS interface within the *precision* of the global timebase.

The computer system of a CS processes Itoms that are received at the CS interface by sensors observing a property of the physical *state* in the entourage of the CS, or are received by messages. Further a CS generates new Itoms that can be implemented as influences on the physical state in the entourage by actuators, or sent as messages into cyber-space, possibly to other CSs.

In summary, a CS implements a time-aware computational element that operates on Itoms which it exchanges with its environment according to its interface specification.

Environment

A CS interacts with two kinds of environments at its interface: cyber space which allows message-based communication and the physical environment which enables physical interactions among CSs.

Cyber Space

Cyber space is a distributed information processing system that enables message-based interactions among CSs by means of direct and indirect cyber *channels*. For instance, the IP-based Internet is a prominent example of a planet-scale cyber space where in principle any two systems with a unique IP address can establish cyber channels and exchange Itoms. The concept of (physical) proximity and neighborhood of CSs does not necessarily play a significant role in cyber space. The physical distance between two CSs exchanging messages via cyber space might only affect the delay of the message, but has no other impact on the communication (e.g., does not change message contents in the absence of *faults*).

A direct cyber channel conveys messages from one sending CS to one or more receiving CS without any modifications.

An indirect cyber channel is established over the state of a shared memory which is located somewhere in cyber space. The sending CSs modify the shared memory by means of state messages. The shared memory is also possibly affected by cyber dynamics a form of *environmental dynamics*. Cyber dynamics are autonomous processes inherent within the cyber space and/or cyber interactions of other not explicitly modeled systems. Finally, receiving CSs obtain the (partial) state of the shared memory. For example, the publish-subscribe [13] communication paradigm is supported by indirect cyber channels where cyber-dynamics (that are in this example part of the system architecture) take care about queueing published messages and notifying subscribers. Many popular message-based middleware platforms support publish-subscribe, e.g., the Robot Operating System (ROS) [34], or Eclipse paho[1], based on MQTT [5].

Physical Environment

The physical environment consists of things and physical fields (energy) whose properties we model as a dynamic network of physical state variables. Such a network of *state variables* can be described in an *environmental model* (for example, see [21] about an ontology-based environmental model) which captures the interrelationships (e.g., transfer delays, functional dependencies, location) of the state variables. For example, the temperature and the pressure of air are physically related and if one is affected by an actuation, so is the other one. Our networked view on the relations of physical state variables allows for a simple composition of environmental models, the consideration of different levels of detail, and taking interfacing effects into account. Note that one can reduce this network to a single set of physical state variables where all relations among the variables need to be described explicitly.

In the physical environment the concept of proximity is essential, because many physical interactions depend on distance (e.g., force fields). In case the entourages of two or more CSs overlap during the IoD, a stigmergic information flow, i.e., a physical interaction, can take place. Overlapping entourages help to limit the size of the environmental model that needs to be considered for the interaction of CSs. Figure 1 shows the network of physical state variables which is under the influence of *environmental dynamics*, but also part of the entourage of multiple CSs. Environmental dynamics are the time-sensitive effects of autonomous processes occurring in the environment.

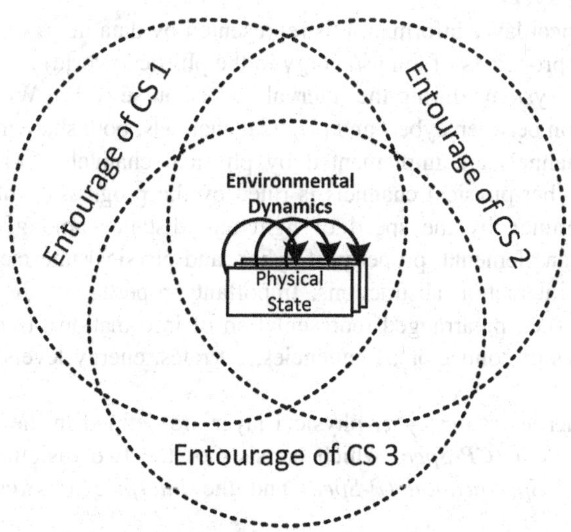

Fig. 1. Overlapping entourage of CPSs enabling physical interaction

[1] https://eclipse.org/paho/.

An actuator is an interface device of a CS which allows the CS to apply changes to one or more state variables of the physical environment that is currently part of the CS's entourage. Besides this actuation also other environmental dynamics may act on the network of state variables. For example, a heating actuator might increase the state variable 'room temperature', while environmental dynamics (heat dissipation) additionally affect the state variable over time. Concerning our environmental model, actuators are connected to the environmental model such that their influences affect the state variables appropriately by considering their placement, actuation delays, and effect propagation through the environmental model.

Sensors within the interface of a CS can observe a state variable of the physical environment. Usually these observations are limited with respect to measurement resolution, temporal *accuracy*, and rate limits. Consequently, such an observation is only partial and noisy. Similar to actuators, also the sensors need to be appropriately connected with the environmental model in order to take their placement and their capability to make observations (measurement delay) into account.

3.2 Interface Layers

Interface layers allow the discussion of system *interface properties* and their definition in *interface specifications* at different abstraction levels and modeling viewpoints. In the following, three interface layers are introduced: the cyber-physical, the informational, and the service layer. The informational layer is an abstraction over the cyber-physical one, while the service layer structures the behavior of a system in a set of capabilities.

Cyber-Physical Layer

At the cyber-physical layer information is represented by data items (e.g., a bit-pattern in cyber space, or properties of things/energy in the physical world) that are transferred among interacting systems during the Interval of Discourse (IoD). While in this layer there is a distinction between cyber- and physical channels, both share many properties, because cyber channels are implemented by physical channels. Consequently, any interaction over cyber-physical channels is ruled by the progression of time and fundamentally constrained by the speed of light and distance among communicating systems. Time is an elemental property of cyber- and physical interfaces and must be considered at all interaction abstractions. Important properties of the cyber-physical layer are: *signals* (i.e., prearranged representation of information), transmission medium, characteristics of connectors, frequencies, bit rates, energy levels.

Interface properties at the cyber-physical layer are defined in the *Cyber-Physical Interface Specification (CP-Spec)* which consists of the two disjoint specifications: *Interface Physical Specification (P-Spec)* and the *Interface Message Specification (M-Spec)*.

Physical Interfaces

Physical interfaces of CSs are realized by energy transformers that are able to *(1)* take observations from the physical environment, and *(2)* set actions initiated from the computer system of the CS in the physical environment. An observation in time-aware CPSoSs is a time-stamped measurement of a physical state variable (a property of a thing). A sensor is an interface device that measures the physical environment and produces observations in the form of digital data (a bit pattern), whereas the sensor design determines which property of the physical environment is observed. Sensor-fusion and state estimation [22] are well researched techniques to improve the fidelity of sensor observations. An actuator is an interface device that accepts digital data and control information (e.g., an actuation deadline) from an interface component, and realizes the intended effect in the physical environment (influences physical state variables).

As physical interfaces enable the interaction with the time-sensitive physical environment, they are time-sensitive as well. Hence, sensor/actuator *latency* and *jitter* affect the temporal accuracy of an observation or the timely effect in the physical environment.

The design of sensors and actuators as well as their placement in the physical environment determines the semantics of the digital in-, and/or, output bit-pattern, i.e., effectively form a basic Itom that contains the observation or actuation. In regard to information theory, we often have that some property of a thing (e.g., voltage level) has additional meaning to its receivers (e.g., digital zero and one), while the actual property of the thing becomes irrelevant, after the measured value has been properly abstracted. This refinement process takes place according to a receiver/sender shared conceptual context. It removes intrinsic information and produced a higher level Itom that contains only the extrinsic information. In computer systems, such overlay-meaning needs to be added as meta data until an Itom is formed which the target CSs can interpret and access correctly. The refinement process also removes unnecessary data that is not needed to convey the intended information, i.e., transforms *raw object data* to *refined object data*.

Take for example a speed limit sign at the side of the road which should be interpreted as such by an autonomous car CS. First, a camera sensor of the CS produces a bitmap Itom where the road side including the speed limit is contained as a large array of pixels. Then, for instance machine learning-based methods take this bitmap Itom, segment the image, and finally extract an Itom describing the speed limit sign. At this point the bitmap Itom including its large object data becomes irrelevant and can be removed from the computer system memory. The new Itom is then further contextualized with surrounding/implicit information (e.g., which metric or non-metric system the country where the road is located uses). Finally, it is possible to construct the speed limit Itom consisting of object data that represents a numeric value and meta data which explains (e.g., decimal, km/h, speed limit) the object data to be usable by the CS.

The Interface Physical Specification (P-Spec) describes the properties of sensors and actuators (e.g., sample rates, value/time uncertainties, observation granularity) in order to exchange Itoms with thc physical cnvironment according to a specified purpose. On the input side the P-Spec specifies the formation of basic level Itoms from sensor observations. On the output side the P-Spec defines how basic level Itoms are implemented by actuators as influences on physical state variables. The P-Spec together with an environmental model allows for the description of stigmergic channels.

Cyber Interfaces
Cyber interfaces produce and/or consume messages, i.e., bit-patterns in cyber space (e.g., an email) according to the Interface Message Specification (M-Spec). The M-Spec consists of three parts [25]: *(1)* the *transport specification*, *(2)* the *syntactic specification*, and *(3)* the *semantic specification*. The transport specification describes all properties of a message that are needed by the communication system to correctly deliver the message from the sender to the receiver(s). A correctly transported message adheres to all temporal and dependability specifications. The cyber interface consists of ports (channel endpoints) where messages are placed for sending, or received messages are read from. A port has the following properties:

- **Direction:** Each port has either the direction incoming (messages can be read from the port), or the direction outgoing (messages can be written to the port).
- **Size:** The size of the data contained in the message determines the port size.
- **Type:** The port type specifies whether the message contains state data and should adhere to a *read/write paradigm* or *event* data and should adhere to the *consume/produce paradigm*.
- **Temporal Properties:** The temporal properties determine the temporal behavior of a message with respect to maximum/minimum delay, maximum *jitter*, periodicity, and bounds on send and receive instants.
- **Dependability Properties:** The dependability properties specify dependability parameters (e.g., reliability, security, availability) of the message transport.

The named syntactic units of a message are called *message variables* [25] and are defined in the syntactic specification. Additionally, the semantic specification links the name of message variable to its explanation, i.e., syntactic and semantic specification define the Itom contained in a cyber message.

For cyber interfaces we can differentiate among several types of *compatibility*:

- **Context compatibility:** The same data (bit pattern) is explained in the same way at the sender and at the receiver.
- **Context incompatibility:** The same data (bit pattern) is explained differently at the sender and at the receiver.
- **Syntactic Compatibility:** The syntactic chunks sent by the sender are received by the receiver without any modification.
- **Full Compatibility:** The Itom that is sent by the sender is received by the receiver without modification.

In case of context compatibility, syntactic compatibility suffices to realize full compatibility. In case of context incompatibility a *gateway* is required to translate the data representation of the sender to a data representation that is compatible with the context of the receiver.

Informational Layer
This interface layer concerns the timely exchange of Itoms by unidirectional channels across interfaces. It provides an abstraction over cyber-physical channels to context-independent [27], direct and indirect information flows among systems and

their environment [28]. The abstraction over cyber-physical channels removes any lower-level details of the interactions that is not relevant for describing the information processing behavior of CSs. Itoms at this layer are maximally refined and explicitly specified, i.e., their meta data is available to the extent necessary for all CSs that are possibly involved with these Itoms. Their realization at the lower-level cyber-physical layer must adhere to the semantics specified at the informational layer, otherwise the abstraction is invalid and there is risk of *property mismatch* among interacting CSs. All for the CPSoS service relevant cyber-physical interactions must be taken into account at the informational layer. Otherwise there are *hidden channels* at the informational layer which might compromise security, safety, or may lead to unexpected behavioral detrimental emergence.

Further, the informational layer focuses on modeling the direct and indirect communication among CSs. There are cases where it is beneficial not to model every system involved in an interaction explicitly, but regard them as anonymous common environment of a smaller set of systems of interest which indirectly interact over this common environment. Indirect communication also allows for decentralized coordination of systems [36] and the description of cascading effects [16].

- **Direct Communication:** Itoms are transferred directly and unmodified from one sending to one or more receiving CSs. Consequently, we model a direct channel by a system that simply forwards Itoms from its input to its outputs according to given temporal and dependability properties.
- **Indirect Communication:** The Itoms of a sending CS affects the state of the common environment of one or more CSs. Additionally, the state of this common environment is possibly affected by environmental dynamics, i.e., time-sensitive processes that act autonomously and independently of the explicitly modeled systems. Finally, receiving CSs read Itoms from the common environment by taking observations. The received Itoms represent a superposition of all influences carried out by other CSs or environmental dynamics. In contrast to direct communication, not all CSs participating in an interaction need to be modeled explicitly, as long as their effects are appropriately considered in the model of the environmental dynamics. We model an indirect channel by instantiating an additional Environmental CS (ECS) which incorporates the behavior of the common environment of indirectly interacting CSs.

An Itom channel is characterized by what kind of Itoms the channel can transport, the sender, one or more recipients, temporal properties, and dependability properties. The *Interface Itom Specification (I-Spec)* describes the Itoms exchanged at the system interface, independently of how the information transfer is actually realized. For example: a system 'car' notifies cars behind about its sudden change of velocity to an immediate stop by 'emergency brake' Itoms. In the cyber-physical layer these Itoms might be implemented by: *(1)* a stigmergic channel between the braking car and the cars behind who observe that the car in front suddenly slows down, *(2)* a stigmergic channel realized by the brake light of the sender and the human operators of the cars behind, and *(3)* a wireless car2car cyber channel between the braking car and the cars behind.

Service Layer
At the service layer, the interface exposes the system behavior structured as capabilities. In contrast to the informational layer, Itom channels are not individually described at the service layer, but only the interdependencies between the exchanged Itoms are specified. If a system with a need is matched with a system that offers the needed capability, the interdependencies must be resolved in the information interface layer with concrete Itom channels. Hence, at the service interface layer there is an instantiable collection of Itom channels per offered capability where generic properties of the Itom channels and their interaction pattern are described.

Systems may provide many services through their interfaces provided that their internal structure can rely on required services. This concept is a fundamental principle in the Service-oriented Architecture (SoA) [12] where components in need of capabilities and components that offer capabilities are brought together by means of a *service registry*, *service discovery*, and *service composition*. A *service provider* is a component that provides a service, while a *service consumer* is a component that uses a service. The service registry is a repository of *Interface Service Specifications (S-Specs)* of capabilities that can be provided by a service provider. Service discovery is the process where service consumers match their service requirements against the available S-Specs in a service registry. Finally, service composition is the integration of multiple services into a new service. The benefits of this service-based view are twofold:

First, there is an immediate reduction of complexity, because one does not need to regard component relations on the basis of single Itom channels anymore. Service consumers can discover services they depend on, and a scheduler can instantiate the necessary unidirectional Itom channels automatically. Further service composition enables the formation of higher-level services based on low-level services. Take the example of a service that provides a humanoid robot the capability to open doors. Such a service would need to implement planning and re-planning of the complex movements realized by lower-level actuation services while constantly taking into account observations (e.g., position of arms, state of the door) from lower-level sensing services.

Second, the coupling of components (integrated in one system) is loose, because the actual constituents of composed services are unimportant background details in the service-based view. This freedom in service composition allows for self-organized system reconfiguration such that the system is able to perform optimally in case new services become available and previously active services become un-available. For example, a service consumer does not depend on a single component to provide the required service, but the service consumer can lookup multiple suitable services from the service registry and choose the optimal regarding computational, communicational, or other costs.

These benefits have been originally observed in the context of free market economy where trading among buyers and sellers led to efficiency in the production and distribution of products.

An S-Spec also includes a set of *quality metrics* that are available for an independent observer to determine the quality of a provided service. Based on these quality metrics, service providers can offer their service under a *Service Level Agreement (SLA)* which consists of *Service Level Objectives (SLOs)* together with the price of a service, and compensation actions in case an SLO of a committed service was not achieved. An SLO describes a quantifiable service objective based on measurable *quality metrics* that can be monitored independently of the service provider. Service providers can publish their SLA with a reference to the S-Spec of an offered service at the service registry, such that prospective service consumers can find and choose an appropriate service provider.

3.3 Interfaces of a Constituent System

Interfaces within Constituent Systems (CSs) that are not exposed to other CSs or the CS's environment are called *internal interfaces*. A CS is embedded in its environment by its *external interfaces*. When applying the principle of separation of concerns, there are three subtypes of external interfaces: *Time-Synchronization Interface (TSI)*, Relied Upon Interface (RUI), and *utility interfaces*. The TSI enables external time-synchronization to establish a global timebase for realizing time-aware CPSoS. Most important for the integration of a CS in a CPSoS is its RUI which is the interface the emergent and operational CPSoS service relies upon. The optional utility interface is an interface of a CS that does not need to be considered for the operational service of CPSoSs.

The purposes of the utility interfaces are to *(1)* configure and update the CS, *(2)* diagnose the CS, and *(3)* let the CS interact with its remaining local environment which is unrelated to the operative service of the CPSoS. These three purposes justify the introduction of the following utility interfaces: *Configuration Interface (C-Interface)*, *Diagnostic Interface (D-Interface)*, and *Local I/O Interface (L-Interface)*. Figure 2 shows all external interfaces of a CS.

In time-aware CPSoSs, the CSs have access to a synchronized global time base with bounded precision. Such a global time base can be established by external clock synchronization over the TSI to, for example, a Global Navigation Satellite System (GNSS) like GPS. Time-awareness allows for temporally ordering observed events and temporally correctly executing timely available actions in a distributed setting. Naturally, in case the communication or computation subsystem or both fail to deliver or execute an action at its deadline, the execution cannot be guaranteed to be temporally correct. However, in a time-aware CPSoS the *temporal order* of observed events – no matter which CS observed them – can be always determined.

We briefly discuss the utility interfaces. The C-Interface is an interface of a CS that is used for the integration of the CS into a CPSoS and the reconfiguration of the CS's RUIs while integrated in a CPSoS. In time-aware CPSoSs the C-Interface allows to update the interface specification of RUIs to realize time-controlled evolution (see Sect. 5.3). A predefined *validity instant* which is part of the interface specification determines when all affected CSs need to use the updated RUI specification and abandon the old RUI specification. This validity instant should be chosen appropriately far in the future (e.g., in the order of the update or maintenance cycle of all impacted CSs). Service providers guarantee that the old interface specification remains active until the validity instant such that service consumers can rely on them up to the reconfiguration instant. The D-Interface is an interface that exposes the internals of a CS for the purpose of diagnosis. Finally, the L-Interface is an interface that allows a CS

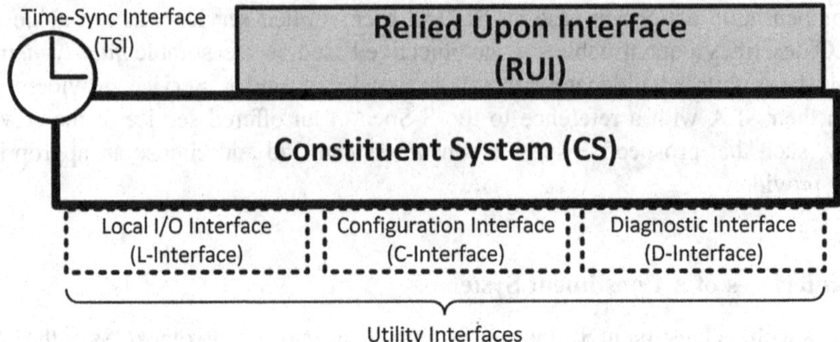

Fig. 2. Interfaces of a Constituent System (CS)

to interact with its surrounding physical reality that is not accessible over any other external interface, for example to realize Human Machine Interfaces (HMIs), or provide other CS-local only services.

A *connected interface* is an interface that is connected to at least one other interface by a channel. Some external interfaces are always connected with respect to the currently active operational mode of a correct system. A disconnected external interface might be the cause of a fault [4] (e.g., a loose cable that was supposed to connect a joystick to a flight controller) which might even lead to a catastrophic failure.

CSs may connect their RUIs according to a *RUI connecting strategy* that searches for and connects to RUIs of other CSs. The RUI connecting strategy is a part of the interface specification of RUIs and searches for desired, with respect to connections available, and compatible RUIs of other CSs and connects them until they either become undesirable, unavailable, or incompatible. For instance, in the global Automated Teller Machine (ATM) network, a cardholder together with a smartcard based payment card form a CS that is most of the time disconnected from any other CSs. The RUI connecting strategy of the payment card CS is influenced by the cardholder's need for cash (desire), nearby located and operational ATM terminals (availability) and whether the ATM terminal accepts the payment card (compatibility).

4 Relied Upon Interfaces

This section discusses the Relied Upon Interface (RUI) model at the previously introduced interface layers, also showing how interface layers are connected. Then the section proposes appropriate execution semantics of the RUI model at the informational layer and closes with a brief discussion of how CPSoS dynamicity is handled by the RUI specification.

4.1 RUI Model Overview

The Relied Upon Interface (RUI) establishes a system boundary of a CS by separating it from its environment. The part of the CS behavior which the CPSoS service relies upon can be observed at the RUI of the CS. Consequently, the interface specification of a CS's RUI hides the possibly complex internal behavior of a CS from the overall CPSoS. However, even more importantly the complexity of the overall behavior of a possibly enormous CPSoS is also hidden from a CS at its RUI. Hence, the RUI specification can be regarded as a complexity firewall because it regulates all interactions taking place across the specified interface. Innate to RUIs, i.e., the points of interactions of CSs, is the transfer of information occurring over these interfaces. It follows an examination of RUIs for each of the three interface layers that we introduced in Sect. 3.2.

RUI Cyber-Physical Layer
Figure 3 gives an overview of cyber-physical interactions at the RUIs of two CSs that are externally time-synchronized and have access to a global timebase. The RUI consists of two sub-interfaces: the *Relied Upon Message Interface (RUMI)* a cyber interface, and the *Relied Upon Physical Interface (RUPI).*

The RUPI consists of sensors and actuators that take and time-stamp observations of and/or act at a defined deadline on some physical state (e.g., the temperature of a room) in the physical environment according to their design. Environmental dynamics (e.g., heat dissipation through walls) act additionally to other CSs on the physical state. CSs that interact with each other over a common physical environment establish a stigmergic channel [28], i.e., they communicate indirectly by influencing and measuring the physical state.

The RUMI allows *(1)* for the unidirectional transport of state and event messages [25] by means of conventional direct cyber channels, and *(2)* for the indirect coordination with other CSs by means of indirect cyber channels. A state message contains only state observations, i.e., the observed state (e.g., temperature of a room) at a specific instant.

RUI Informational Layer
The informational layer abstracts over informational context-sensitivity, and focuses on direct and indirect information flows among CSs. An indirect channel (cyber or stigmergic) is modelled by instantiating an additional Environmental CS (ECS).

This interface layer is useful during the design of CPSoSs (e.g., model-based design, design space exploration), as well as in the analysis of CPSoSs. For example, we believe that identifying causally related interactions among CSs is paramount for detecting and predicting emergence (see Chap. 3). Naturally, for finding such causal relationships the RUI specifications, the associated interface models, and the environmental models need to be accurate regarding reality, i.e., there should not be any hidden channels. A hidden channel is a latent information flow among CSs that has not been considered by the modeler. Hidden channels might close feedback loops that are believed to be liable for possibly undesired detrimental emergence [28]. In case some

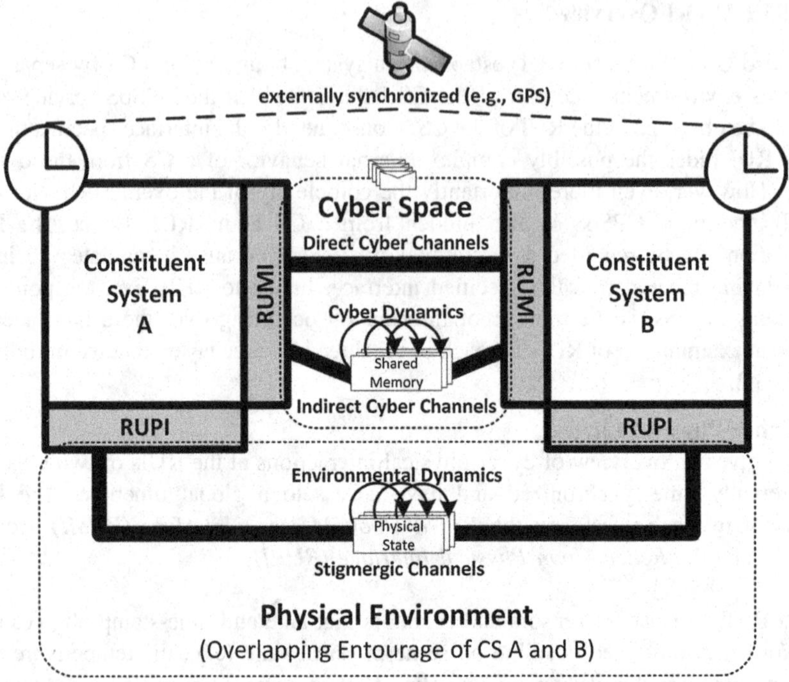

Fig. 3. Relied Upon Interfaces (RUIs) at the cyber-physical layer

behavior is observed at the RUIs of CSs during CPSoS operation, but cannot be reproduced in a simulation at the informational layer, there are hidden channels present that should be identified.

RUI Service Layer

At the service interface layer, we introduce *Relied Upon Services (RUSs)* that are provided at the RUI of a CS. They are described in the Service Specification (S-Spec) of the RUI as a set of RUS-related operations. A service operation is a behavioral abstraction over one or more unidirectional Itom channels. It groups them together and defines their interaction pattern, i.e., the sequence of all operation-related Itoms over all channel endpoints from the perspective of the service provider. Examples of interaction patterns are: request-response, notify, or solicit. Actual Itom channels, or consequently cyber-physical channels are only instantiated (or their provisioning considered) if a RUS is committed to a service requester.

Besides defining the operations of a RUS, the S-Spec also includes a set of quality metrics that allow an independent observer (e.g., a monitoring CS) to determine the quality of a RUS provided at a CS. Based on this quality metrics, a RUS provider can publish its Service Level Agreement (SLA) at the service registry such that service requesters are able to find and request suitable services. RUS providers and consumers are CSs. The service registry – depending on dynamicity and business requirements of the particular CPSoS – can be either realized as another CS (operated by an SoS authority) to allow for a runtime RUS composition, or it is realized in an off-line manner.

At the service level we model the emergent CPSoS service as a set of dependencies on the required RUSs, such that any CS that wants to use or benefit from the emergent CPSoS service needs to provide these RUSs. However, a CS does not need to directly provide all or even any RUSs a given emergent CPSoS service depends on, as long as the CS is able to request and consume them from other RUS providers.

Example

This section shows in a small example how the interface layers of the RUI are connected. The example CPSoS consists of n interacting CSs. At the cyber-physical interface layer, it contains CSs that interact by using direct and indirect cyber-physical channels. In the informational layer these channels correspond to Itom channels and Itom processing subsystems that implement the behavior of indirect communication. Finally, at the service layer we are able to group channels that are associated with a service and express service dependency relationships.

Cyber-Physical Layer

Figure 4 shows cyber-physical interactions realized by some concrete technology (e.g., data exchanged in cyber space by a TCP/IP network stack, physical location of the CS on a street influenced by actuators). To relate the channels among the CSs and their environment across all interface layers, we draw all channels related to a distinguishable service with the same style. Cyber Channels (CCs) are drawn with a solid line style, while Physical Channels (PCs) are drawn with a dashed line style. Some of the CCs and some of the PCs are labeled for easier identification.

CC 1 and CC 2 are direct cyber channels of the same service (e.g., a database lookup service realized by a request and a response channel). CC 3 and CCs originating from CSs 3 to n-1 are writers of an indirect channel. For example, they publish information whether an alarm occurred. This indirect channel has only one reader (CC 4): CS n which could be an alarm monitor. Further, there is a stigmergic channel realized by PC 1 (actuator which is part of the CS 1 RUPI), physical state variables, and PC 2 (a sensor device of the CS 2 RUPI). Another stigmergic channel (realized in the figure via the overlapping entourages in the lower right of the physical environment) is shown where all CSs are able to influence physical state variables and also observe them, for example the position of CSs on a street.

Informational Layer

Figure 5 shows the example CPSoS at the informational layer. All direct CCs have a corresponding Itom Channel (IC), for example see IC 1 corresponds to CC 1. The indirect CC is realized by an additional Environmental CS (ECS) which implements the behaviour of the indirect CC described by an appropriate environmental model (shared memories with all acting cyber dynamics). Also for each stigmergic channel an additional ECS realizes the environmental model (environmental model with all acting environmental dynamics).

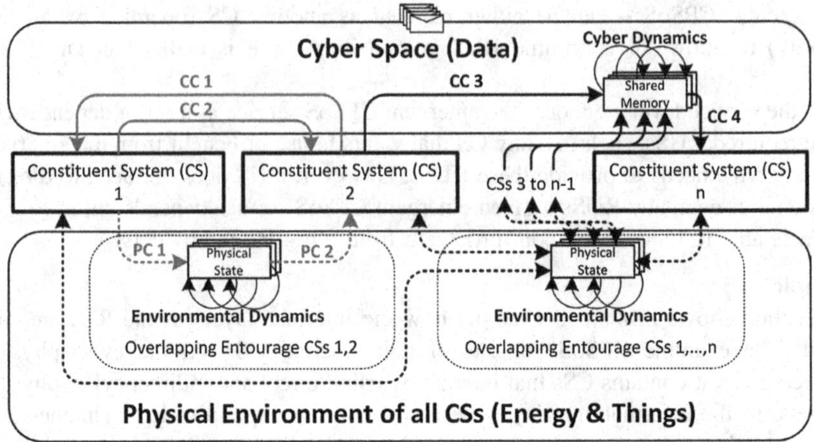

Fig. 4. Constituent Systems (CSs), Cyber Channels (CCs), and Physical Channels (PCs) at the cyber-physical interface layer

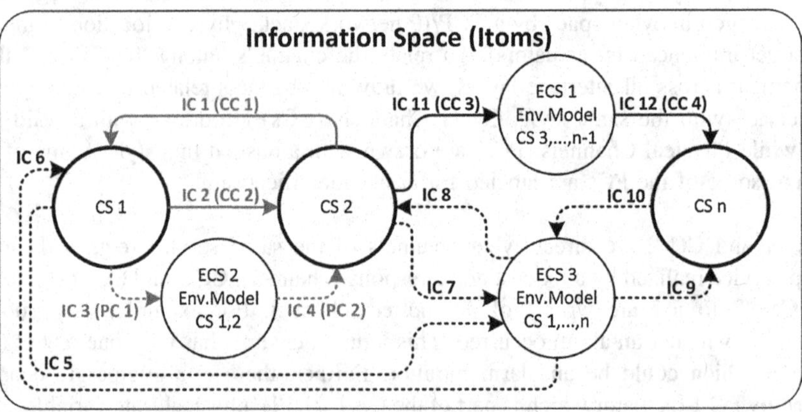

Fig. 5. Constituent Systems (CSs) and Itom Channel (ICs) at the informational interface layer

Service Layer

The service layer of the example CPSoS consists of four services in the dependency relation shown in Fig. 6. For example, service A depends on the environmental services C and D. Service A might be a database lookup service provided by CS 2. The incoming and outgoing arrows on the left of the service vertex symbolize the two Itom channel ports from the perspective of the service provider (one input port and one output port). The three environmental services B, C, and D are provided by ECSs (e.g., environmental service D is provided by ECS 3 in the informational layer). CSs that consume such services need to either act as an influence/writer, or as an observer/reader. The ports on the left side of an environmental service represents the influence/writer service consumer

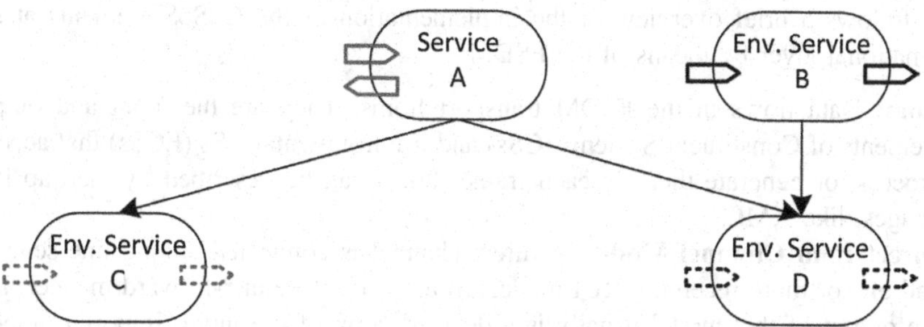

Fig. 6. Service interface layer

(e.g., CS 1 is an influence/writer of ECS 2), and the ports on the right side of an environmental service stand for the observer/reader service consumer (for instance, CS 2 is an observer/reader of ECS 2).

4.2 Execution Semantics of the Informational RUI Model

The interface model describes the part of the system behavior which is observable at that interface. In the previous section we presented an overview of the RUI model. In this section we want to enrich our interface design by suggesting a Frame-based Synchronous Dataflow Model (FSDM) as an execution semantic for the informational layer. Having such execution semantics, allows the detailed study of CPSoSs with respect to behavioral properties at the informational layer in the value domain and in the temporal domain.

Frame-Based Synchronous Dataflow Model
The FSDM is prominent in modeling dependable real-time systems that interact with physical systems or models of them. Further, there are many high-quality tools and languages to design, execute/simulate, and verify such synchronous models, e.g., GIOTTO [20], and Lustre [18].

In the FSDM the dense physical time is discretized into *frames*, i.e., periodic sequences of constant duration. Each of the frames consists of a synchronization phase and a processing phase. During the synchronization phase the input is received (in a sample and hold manner) from the system environment and the output is sent to the environment. In the processing phase the system's function calculates the next state and the output from the input and the current state. Only during the synchronization phase there is interaction between a CS and its otherwise free-running environment. Under the synchronous hypothesis [33] a frame duration is short enough such that the system appropriately reacts to changes in its environment. Hence, the frame duration is determined by the environmental dynamics. For instance, for a keyboard-based Human Machine Interface (HMI) a frame duration of 50 ms is appropriate for most applications, while for a crash detection system in a car a frame duration lower than 1 ms is more appropriate.

It follows a brief overview of the implementation of the CPSoS elements at the informational layer by means of the FSDM:

- **Itom:** Data flows in the FSDM transport Itoms. They are the input and output elements of Constituent Systems (CSs) and Environmental CSs (ECSs) that access, process, or generate them in each frame. Itoms can be described by markup languages, like XML.
- **Direct Itom Channel Model:** A direct channel is connected among one sending and one or more receiving RUI models. It acts in a store-and-forward manner, i.e., the output of this model usually is a delayed copy of the input. Important model attributes are delay, jitter, and optional fault behavior.
- **CS and ECS RUI Models:** RUI models can connect to channel models according to their connecting strategy (see Sects. 3.3 and 4.3). In time-aware CPSoS the internal state of a CS includes the current global time on which input, output and computation actions can be based. ECS RUI models describe the behavior of the common environment of indirectly interacting CSs. They need to integrate the Itoms received from connected CS RUI models in their internal state and apply specified environmental dynamics to this internal state at each frame. The output of ECSs reflects their internal state according to the (often limited) observation capabilities of the receiving connected RUI models. ECSs that model large state spaces and computational intensive environmental dynamics of the physical environment can limit the considered interactions to the overlapping entourages of the involved CSs.

Implementation Considerations
In time-aware CPSoSs the CSs as well as the cyber channels might not be able to guarantee the reaction time constraints imposed by the FSDM. Still, if inputs (and outputs) adhere to the state semantics, there are sophisticated state estimation techniques [22] available to tolerate occasional violations of the synchronous hypothesis (even to the extent of incorporating state observations with significant delay and jitter, e.g., cf. [11], Fig. 2).

In time-aware CPSoSs the CSs can use the available global timebase to drive a periodic control subsystem of a CS as follows: Frame start instants are aligned with a *tick* event of the global time and frame durations are multiples of the *granularity* of the global time. Further, the periodic control subsystem is responsible in the processing phase to activate application tasks, and during the synchronization phase to conduct send and receive actions. There is implementation technology available that readily supports the FSDM. Examples are: TTEthernet [35], TTP/C, or the ACROSS Multi-Processor System-on-Chip [37].

4.3 RUIs Under Dynamicity

Dynamicity describes the reaction or reconfiguration capabilities that have been already considered in the CPSoS design. Therefore, any supported dynamicity needs to be defined in the RUI specifications. We examine three prototypical dynamicity cases:

- dynamicity with respect to connecting two or more CSs at their RUIs,
- dynamicity in making (partial) CS or emergent CPSoS services available to other CSs (RUS composition), and
- dynamicity to adapt to changes in the environment.

RUIs might be connected only for a finite duration within an Interval of Discourse (IoD). A disconnected RUI might be a normal, fault-free interface state and may result at most in CS service degradation, but not in CS failure. This key aspect of RUIs is responsible for the impossibility to establish a static system boundary of a CPSoS. CSs may disconnect and reconnect and they might even be part of multiple CPSoSs (e.g., a modern day NFC enabled smartphone can be part of the global telephone network, the Internet, and the global ATM network which are three large and independently operating CPSoSs).

The RUI connection strategy is part of the interface specification of RUIs that regulates how CSs establish connections. All RUI connecting strategies are local to their respective CS. Still, they influence the dynamic global network topology of a CPSoSs. Consequently, they are co-responsible for the occurrence and regulation of self-organization and emergent phenomena.

At the cyber-physical and informational layers of RUIs it is cognitively complex to describe the dynamic CS interaction that is necessary for accessing a service on the basis of individual channels, because the specific client CSs are unknown before runtime. For example, take a CS that offers a database service. This database service requires a request and a response Itom channel per client CS. For n client CSs one would need to specify 2n dedicated channels at the cyber-physical or informational interface layer. When considering the same situation at the service layer, only the database service together with the two required channels needs to be specified. The mechanisms of service discovery and service composition allow a scheduler to automatically instantiate the request and response channels for each client CS during runtime.

Finally, CSs might need to react to the changing environment and need to reconfigure the set of offered services in order to: *(1)* accomplish overall CPSoS goals (e.g., limit total energy budget, enter a safe state, …), or *(2)* tolerate faults (e.g., suddenly disconnected CSs or failing CSs). At the service interface layer, paradigms known from the Service-oriented Architecture (SoA) are readily available for use. For example, one such paradigm is to replace services with a degraded version, or to replace failed services altogether with services from other (redundant) CSs [21].

5 Interfaces in Evolving Cyber-Physical Systems-of-Systems

Evolution of Cyber-Physical Systems-of-Systems (CPSoSs) concerns design modifications introduced into the interacting Constituent Systems (CSs) that are triggered by changes in the CPSoS environment. Changes of the CPSoS environment might include, for example, advances in technology, or are changes in societal or business needs. Often these needs originate from the desire to change a service towards increased efficiency or the wish to introduce new services altogether. Ultimately, evolutionary changes to the design and consequently operation of the CPSoS should counteract obsolescence in order to keep the CPSoS relevant, increase its business value for involved stake holders, while not deteriorating already provided and still needed services.

When discussing evolution in possibly large and complex systems like CPSoSs we distinguish between *unmanaged* and *managed evolution* [32]. In unmanaged evolution there is no guidance about how a CPSoS evolves. Owners of CSs are free to change services and cooperation with other CSs is motivated by each CS owner's own gain in perceived business value. Facebook, Wikipedia, Google services, and Twitter messaging are examples of CSs whose interfaces are controlled and evolved by the respective owning companies. The composition of such CSs is not driven by a specific central purpose and leads to *virtual SoSs* (e.g., Twitter/Facebook integration of fitness tracking and training applications, or a clever integration of Wikipedia and Facebook to realize file-sharing services). Consequently, unmanaged evolution is most suitable for virtual SoSs where there is no clear central purpose.

In this section we focus on the technical realization of *managed evolution* [32] which is most appropriate for *collaborative SoSs* (but also *directed* and *acknowledged SoSs*, see Chap. 1 for details about SoS classifications). Managed evolution has been originally suggested in the context of large, long-term operational software systems that also show many similarities with CPSoSs (complex functional, semantical, temporal, technical, and operational interdependencies of interacting systems with non-trivially replaceable legacy subsystems). In collaborative SoSs, managed evolution must be planned and supervised by an SoS authority, i.e., an organizational entity (for example established by a CPSoS consortium or enterprise), such that *"the efficiency of developing and operating the system is preserved or even increased"* [32]. The SoS authority has a specific CPSoS purpose in mind, maintains the specifications of Relied Upon Interfaces (RUIs), and has a set of capabilities in order to manage CPSoS evolution. The capabilities are:

- means to introduce changes into a CPSoS by ultimately modifying RUI specifications of CSs,
- monitor the *evolutionary performance* of the evolving CPSoS, and
- give *incentives* to steer the evolutionary process forward, i.e., influence CSs to implement modified RUI specifications.

In the following we discuss scope and challenges of managed evolution in time-aware CPSoSs and in particular investigate evolution at Relied Upon Interfaces (RUIs) of CSs. We attempt to confine risks associated with unexpected detrimental emergence and finally suggest a set of guidelines how to handle evolution in time-aware CPSoSs.

5.1 Scope and Challenges

Managed evolution as described in [32] discusses and addresses challenges related to large, complex, and long-term operational software systems. The comprehensive findings of the authors apply to collaborative SoSs, because they share all characteristics of the systems discussed in [32]. For example, one important challenge the authors address is maintaining *agility* (i.e., the ability to efficiently implement evolutionary changes) while also carrying out necessary changes to increase the system's *business value*.

Further, the authors extensively examine organizational, governance, and even cultural or people aspects of evolving systems, while in this chapter we mostly concentrate on technical aspects. In the following, we identify challenges specifically occurring in the evolution of time-aware CPSoSs that we want to tackle:

- **Continuous Evolution:** Compared to traditional monolithic systems, CPSoSs need to continuously evolve, because replacement of the overall CPSoS as well as a redesign from scratch (green-lawn or greenfield approach) are infeasible with respect to involved costs, risks, or inacceptable operational discontinuities. For example, in the global Automated Teller Machine (ATM) SoS at one point in time a more secure chip-based payment card has been introduced. It is immediately clear that an instantaneous replacement of all payment cards together with the replacement of all Points of Sale (PoS) and ATM terminals worldwide cannot be done because of scaling issues. Consequently, CPSoSs need to evolve continuously, usually during runtime.
- **Multi-version Evolution:** An unavoidable consequence of continuous evolution is that parts of an CPSoS are at different evolutionary states. In large and long-term operational CPSoSs the CSs cannot be replaced or upgraded simultaneously. Hence CSs need to be able to interact with older versions of themselves. Further, CSs cannot be arbitrarily updated and might become at some point legacy CSs that need wrapping to be able to interact with the further evolved CPSoS. In CPSoSs we have CSs in multiple versions and need to take care about appropriately wrapping legacy CSs.
- **Unexpected Detrimental Emergence:** Evolutionary changes might lead to unexpected emergence that is highly undesired (e.g., compromised security or safety properties of an CPSoS). Unfortunately, unexpected emergence cannot be easily or reliably predicated and may only be discovered by accident, because the boundary of a CPSoS is not static and there might by unforeseen environmental effects. For example, consider the British Airways Flight 38 where a Boeing 777 crashed on January 17th, 2008 shortly before the runway at its destination: both engines suddenly failed during landing. The investigation concluded that ice has formed in the fuel system which restricted the fuel flow to both engines [1]. At that time the formation of icc was an unconsidered environmental effect which was only revealed after an accident occurred.
- **Evolving CPSoS Dynamicity:** AMADEOS CPSoSs feature architectural support for adaptive monitoring, analyzing and planning via cognitive and predictive models, and execution of reaction strategies. These architectural means to handle

CPSoS dynamicity need to evolve together with changes in the CPSoS service. For example, in case an evolutionary change allows more efficient use of crossroads and use of street lanes, different traffic situations might arise which require different reaction strategies in order to optimize traffic flow and minimize detrimental effects of faults.

Note that we do not claim that this list of challenges is complete. There might be more challenges related to evolution, especially, if one wants to discuss CPSoSs of specific application domains.

5.2 Local and Global Evolution

We call changes within a CS that do not affect its interactions with other CSs *local evolution*. Local evolution does not modify any of the CS's RUI specifications and consequently remains invisible at the CPSoS level. Still, local evolution is important to optimize the operation of CSs, introduce new or change local-only CS services, or prepare CSs for a pending global *evolutionary step*.

Local evolution harbors the risk of introducing hidden channels (i.e., unconsidered interactions) among CSs which could lead to emergent effects. Hence, local evolution must carefully respect RUI specifications which forbid – in principle – any interaction of a CS with its environment that is not explicitly defined. However, in praxis this is difficult to achieve, especially in relation to stigmergic interactions over a common physical environment. For example, a processor may leak information via its power consumption. In case some local evolutionary change enables an attacker to measure the power consumption, CPSoS security might be compromised.

In contrast to local evolution, we call changes that affect the interactions of CSs and thus the service of the overall CPSoS *global evolution*. As such, global evolution concerns the change of RUI specifications and how these changes are coming into effect. In CPSoSs we have continuous evolution. Consequently, CPSoSs cannot be changed radically, but need to evolve gradually towards changed or new goals in evolutionary steps of limited scope preferably with predictable effects.

Inspired by biological evolution of life (cf. Darwin and natural selection) we regard CPSoS evolution as a tree-like search towards adaptation to environmental conditions. The search space consists of all possible changes that can be realized to address (changed) environmental conditions. Naturally this search space is too large to explore exhaustively; only some of the possible changes are actually realized and can be represented as a tree-like search space exploration. Figure 7 sketches how we view the process of global evolution in CPSoSs. Each of the vertices of the acyclic graph represents a specific *evolutionary state* or version of the CSs making up a CPSoS. The edges in the graph represent evolutionary steps. The vertical dashed line represents the instant *now* which separates the past (versions of CSs that exist and may or may not be in operation) from the future (versions of CSs that are planned and are not operational yet). We assume that some versions of CSs are compatible at the present. In the figure we have indicated two overlapping sets of versions of CSs that are compatible (upper and lower lassos containing a set of vertices each).

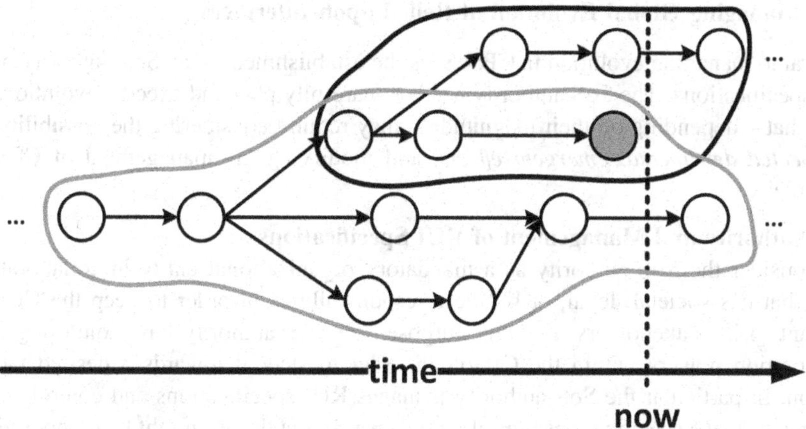

Fig. 7. Tree-like search towards adaptation

Three fundamentally different types of evolutionary steps have been identified:

- **Basic Step:** A basic evolutionary step is a linear, incremental update from one version of a CPSoS to the next one. In Fig. 7 a basic step is represented for example between the two most left vertices.
- **Fork:** In this case two or more different versions evolve from the same ancestor version. For example, a smart grid CPSoS might have evolved in one country up to a certain version which is adopted by a different CPSoS consortium in a different country. Over time these two versions might diverge further up to a point where CSs from one evolved version are not compatible anymore with the CSs from the other evolved version. Figure 7 exemplifies this case after the second vertex from the left where there is a fork into three different versions.
- **Merge:** Finally, two versions of CSs that are part of the same CPSoS might merge in a later version in order to reduce unnecessary functional redundancies, benefit from standardization efforts, or consolidated interfaces. In Fig. 7 this case is depicted in the vertex that has two incoming edges.

Note that in some cases a version can be abandoned and not evolved further as we have illustrated in Fig. 7 by using the shaded version vertex which has no outgoing edge. CSs of such an abandoned version will turn into legacy CSs until they become obsolete.

In terms of managed evolution, forks create mostly business value, while merges increase agility. Basic steps either increase business value or agility. It is in the responsibility of the involved SoS authorities to appropriately control evolutionary steps such that the business value for all stakeholders stays viable while simultaneously agility is not reduced. Both, business value and agility are quality metrics that are composed of CPSoS application-specific quality metrics. For example, the business value of a CPSoS consisting of autonomous cars that should transport humans in a city while minimizing environmental pollution can be assessed by average transportation time per km and average pollution per km.

5.3 Managing Global Evolution at Relied upon Interfaces

Integral to managing evolution in CPSoSs is the establishment of an SoS authority over RUI specifications. The SoS authority needs to carefully plan and execute evolutionary steps that – depending on their magnitude – may require considering the possibility of *unexpected detrimental emergent effects*, and modifying the management of CPSoS dynamicity.

SoS Authority and Management of RUI Specifications
We consider the SoS authority as a mandatory organizational entity in collaborative SoSs that has societal, legal, or business responsibilities in order to keep the CPSoS relevant to its stakeholders. For this purpose, the SoS authority has monitoring and amelioration powers within the CPSoS in order to steer it towards a desired target version. In particular the SoS authority manages RUI specifications and controls how changes of these specifications are rolled out. An SoS authority might be composed of representatives of key CPSoS stakeholders, like CS manufacturers or governments.

SoS authorities select and adopt a suitable set of standards developed by standardization organizations such as the Institute of Electrical and Electronics Engineers (IEEE), the Society of Automotive Engineers (SAE), or the Object Management Group (OMG). The role of standardization organizations in CPSoSs is to provide a stable and broadly accepted conceptual and technological basis for the realization of RUI specifications. For example, the SAE J1708 standard specifies the serial communication (physical layer) of Electronic Control Units (ECUs) in heavy duty vehicles.

In the following we outline a technical realization of RUI specification management on a Service-oriented-Architecture (SoA) approach. *Authorized Relied Upon Service (RUS) specifications (S-Specs)* are administrated at a *service registry* (see Sect. 3.2). Only the SoS authority can authorize and publish S-Specs at the service registry. For example, a CS owner[2] participates in the CPSoS by being a *RUS provider* that offers the RUS in compliance to an authorized S-Spec. In support of multi-version evolution the service registry needs to feature *version management* for authorized S-Specs, i.e., the SoS authority can add new versions of S-Specs to the service registry that coexist with older S-Spec versions. Now owners of CSs that provide RUSs have the possibility to specify in their respective SLAs to which specific version of an authorized S-Spec they refer to. For each supported S-Spec version a different SLA needs to be offered by the CS owner.

Besides managing RUI specifications the SoS authority needs to assess and steer the overall evolutionary process of the CPSoS. The performance of the evolutionary process of a CPSoS is derived from measurable quality metrics describing business value and agility of the CPSoS. Consequently, the measurement of the evolutionary performance can be efficiently integrated in the monitoring and analyzing blocks of the AMADEOS solutions concerning the management of CPSoS dynamicity (see Chap. 7). One example of an important quality metric for the assessment of the

[2] For the sake of brevity we assume here that the owner of a CS is also its manufacturer and user.

evolutionary performance in CPSoSs is the *adoption rate* of existing or newly introduced CSs to new or changed versions of S-Specs.

The adoption rate is controllable by the SoS authority who can give *incentives* in order to move the evolutionary process towards a desired target version. Incentives can be advantages or disadvantages where even monetary penalties apply in case a CS is not upgraded to a more recent version. For example, in the global Automated Teller Machine (ATM) network SoS the payment card industry shifted liability at a specified deadline from money institutes to the Point of Sale (PoS) operators (usually merchants), if they used old and insecure equipment to process payment cards. As the deadline of the liability shift approached and passed, PoS operators risked compensating monetary losses caused by fraud from their own pocket, if an insecure PoS terminal under their responsibility was involved. This strong incentive forced PoS operators to upgrade to more secure PoS terminals where they are not liable in the event of fraud.

Magnitude and Effects of Evolutionary Steps
We define the magnitude of an evolutionary step by considering which of the interface layers (see Sect. 3.2) of RUIs are affected by it:

- **Cyber-Physical Layer:** Technological advances (e.g., different communication protocols, more energy efficient sensors and actuators) may lead to a change of how cyber-physical interactions are carried out. In the case that other interface layers remain unaffected (i.e., there is no change in the information flows) we have a *minor evolutionary step*. A minor evolutionary step does not require any further considerations concerning emergence and managing CPSoS dynamicity. In the context of the AMADEOS Architectural Framework (AF) detailed in Chap. 5, a minor evolutionary step only concerns differences in the implementation level.
- **Informational Layer:** We consider changes at the informational layer as a *major evolutionary step*, because a change in the Itom interactions of CSs needs to be carefully assessed with respect to emergence and the management of CPSoS dynamicity. Regarding the AMADEOS AF a major evolutionary step implies changes at the logical, conceptual or even up to the mission level.
- **Service Layer:** Changes at the service layer are always accompanied by changes in the underlying interface layers. Consequently, a change at the service level also represents a major evolutionary step that has implications on emergence and the management of CPSoS dynamicity.

Methods from Scenario-based Reasoning (SBR) [6] can be employed to pre-evaluate effects of evolutionary steps according to CPSoS-specific quality metrics under quantified uncertainty.

Handling Continuous Evolution
It remains to discuss the transition from one CPSoS version to an evolved version. Continuous evolution in CPSoSs is based on the principle of backward compatibility of its CSs. This backward compatibility needs to be established by upgrading or introducing new CSs that also support the interaction with non-upgraded CSs.

We have already described a multi-version service registry and that Relied Upon Service (RUS) providing CSs can support multiple versions of S-Specs. Now we want to emphasize that also CSs requesting RUS have the possibility to consume different versions of a RUS. In fact, the more versions of RUSs a CS is able to provide or consume, the 'smoother' a CPSoS is able to evolve. Naturally, newly introduced RUSs should not disrupt existing or legacy RUSs.

At some point conflicting or obsolete RUSs need to be retired and old, non-upgradable CSs that depend on them or provide them become incompatible with more recent CSs. In case these old and non-upgradable CSs are still essential for the operation of the CPSoS they become legacy CSs and their service needs to be appropriately wrapped. For example, a wrapping CS can be introduced in an evolutionary step. The wrapping CS is able to offer the service of the legacy CS encapsulated in a version of a RUS that is compatible with all non-legacy CSs.

In time-aware CPSoSs there is the benefit of a global time that allows temporally coordinating the execution of an evolutionary step. The SoS authority can define validity instants in new versions of RUI specifications such that they are switched on at a specific instant. For example, a desired emergent effect may only occur if a critical number of CSs adhere to the new RUS version simultaneously. Also, for some CPSoSs it might be useful to steer its CSs first to a safe, ground, or dormant state, then perform further (physical) upgrades, and finally awake them to interact in the evolved CPSoS (see car-recalls and repair procedures in the automotive domain).

5.4 Avoiding Detrimental Emergence

The occurrence of unexpected detrimental emergence is a problematic case in engineering, operating and evolving CPSoSs. Against our expectations and current predictive capabilities something harmful and possibly catastrophic happened. Why can we not prevent unexpected emergence by design? While the engineering process is indeed responsible for the interaction abilities of the CSs, CPSoSs are also open systems that interact with their environment. This environment of a CPSoS may change over time and/or might be insufficiently understood, i.e., in general we cannot be certain that our models of the CPSoS environment are complete. Also the boundary of a CPSoS is dynamic and influences how the CPSoS environment affects the CPSoS itself. For example, consider a fault-tolerant CPSoS: If the number of its redundant CSs is small, an environment that causes intermittent faults in CSs at a constant rate (e.g., radiation) is much more hostile than compared to the same CPSoS where the number of redundant CSs is large.

An evolving CPSoS attempts to adapt to changes in its environment (real changes or a changed understanding of the environment). Consequently, there are two co-dependent causes that enable the occurrence of unexpected detrimental emergence in CPSoSs:

- An evolutionary step that changes how CSs interact, and
- a change in the CPSoS environment and/or hidden channels.

To the best of our knowledge there is currently no scalable theory to eliminate the first cause with certainty. The number of possible interactions in real-world CPSoSs (physical environment!) diminishes any hope to exhaustively test for all known emergent phenomena, and to evaluate them with respect to their possibly detrimental effects. Even if there was a theory solving this issue, one fundamental problem remains: Identifying unknown emergence, i.e., effects or properties that are conceptually novel at the macro-level (SoS level), but are not present in the non-relational phenomena of the parts at the micro-level (level of CSs).

Unfortunately, the situation is even worse concerning the second presumed cause that enables the occurrence of unexpected emergence. Changes in the CPSoS environment are outside the sphere of control of a CPSoS consortium. Some of these changes may lead to the occurrence of previously rare or unlikely interactions of CSs which may in return result in a (detrimental) unexpected emergent phenomenon. Finally, hidden channels are an unfortunate consequence of our ignorance about the CPSoS environment. They may close causal loops or enable cascading effects which again could trigger (detrimental) unexpected emergence.

In summary, it appears that in principle we cannot prevent all first occurrences of detrimental emergent phenomena with absolute certainty. Further, both an ill-conceived evolutionary step, and (unlucky) changes in the CPSoS environment may lead to undesired emergent phenomena. In the following subsections we discuss a mitigation strategy based on results described in Chap. 3.

Mitigation Strategy

In order to minimize occurrence and subsequent damage due to unexpected detrimental emergence we suggest a mitigation strategy consisting of the following procedures:

- **Augmentation of CPSoS design with expectations about nominal operation:** Relied Upon Service (RUS) specifications should contain assertions that indicate whether the RUS is provided and consumed nominally and according to the designer's expectations. Runtime monitoring implemented in the management of CPSoS dynamicity (see Chap. 7) can check the defined assertions against all interactions of CSs and log as well as timestamp any occurring anomalies.
- **Discovery of the onset of unexpected emergent phenomena:** Quality metrics associated with the onset of emergence (e.g., critical slow-down, density), unexplainable anomalies, and patterns of previously diagnosed and analyzed emergence should lead to the discovery of the onset of (detrimental) unexpected emergence. Again this procedure should be implemented in the management of CPSoS dynamicity.
- **Diagnosis and analysis of unexpected emergence:** After an unexpected emergent phenomenon has been discovered, it must be carefully diagnosed and analyzed. This procedure must reveal the trans-ordinal law and it might expose hidden channels or changes in the CPSoS environment that have not been noticed yet. Based on the result of the analysis inaccurate (environmental) models should be corrected and an appropriate evolutionary step planned to prevent the now expected detrimental emergence.
- **Prevention of detrimental emergence by design:** As soon as new expected detrimental emergence has been found an evolutionary step should be performed

that ameliorates ongoing detrimental emergence and prevents its further occurrence. Amelioration can be implemented in the management of CPSoS dynamicity by deploying suitable reaction strategies (e.g., introduction of randomness to break unintended synchronization [31]). Prevention of detrimental emergence should be implemented by appropriately constraining the RUIs of CSs such that their interactions do not lead to the detrimental emergent effect anymore.

- **Prediction of detrimental emergence:** When planning an evolutionary step, we can predict/search for detrimental emergence by applying analytical methods (e.g., finding causal loops, cascading effects) as well as simulation of models of the CPSoS and its environment.

Detecting Unknown Emergence

Often emergent phenomena are associated with some kind of regularities or shift in densities, but looking for something unknown that does not appear to have any generic and unique characteristics limits our detection abilities. Mogul [31] suggests building a library of signatures of emergence that occur in distributed computer systems and gives interesting examples: trashing caused by ‘unlucky’ scheduling (not just overprovisioning), unintended synchronization, unintended oscillation or periodicity, detectable by spectral analysis, deadlock and livelock, phase change, chaotic behavior.

While building such a library still requires some decision procedure to classify anomalies as emergent, it would – together with monitoring – allow for efficient detection of already encountered emergence. A decision procedure to classify anomalies could be supported by unsupervised machine learning. Self-organizing Maps (SOMs) appear to be a particularly interesting technique where clusters of structure or density differences in data can be found (see Fig. 8, right).

The SOM [23] is an unsupervised machine learning approach that is based on an artificial neural network of a usually low-dimensional topology (e.g., two-dimensional as depicted in Fig. 8, left). During the training process possibly high-dimensional input data is (after appropriate pre-processing [19]) firstly vector quantized and secondly mapped to the SOM while attempting to preserve the original topology of the input data. Various visualizations for SOMs exist to both make the information contained in the SOM accessible for analysis, but also to allow for assessments concerning the quality of the vector quantization and the quality of the topology preservation. Consequently, visualizations are critical for the interpretation of SOMs. Visualizations often focus on a single or a few aspects of the SOM, i.e., for analyzing SOMs, it is often necessary to study multiple visualizations of them in combination. For example, visualizations can express structural information about input vector density relations, distance relations among mapped input vectors (also called topology), or class information. Further, SOMs can be used to both semantically relate different input samples among one another, and to predict the relation of new input samples to input samples used in training.

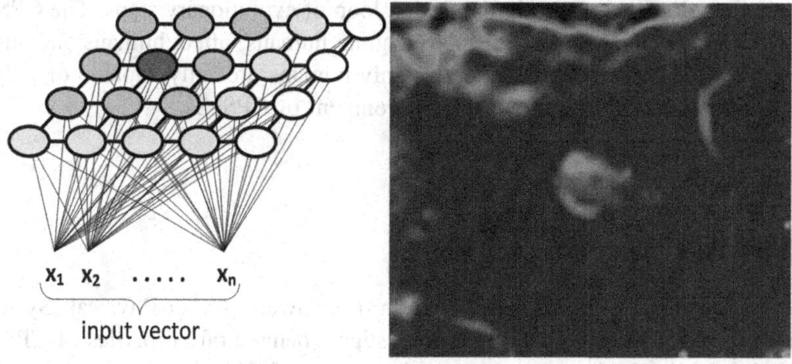

Fig. 8. Self-Organizing Maps (SOMs) reveal emergent regularities in high-dim input data

5.5 Design Guidelines for Evolvable Systems-of-Systems

We conclude this section with a list of design guidelines for evolving CPSoSs that in particular apply to collaborative SoSs:

- Precisely specify temporal properties of RUIs. Local evolution might inadvertently violate RUI specifications, particularly in the temporal domain. Such violations may lead to hidden channels enabling undesired interactions among CSs. The undesired interactions possibly cause unexpected detrimental emergence. Therefore, CSs need to be checked (e.g., by the CPSoS dynamicity management) concerning their conformance to authorized RUI specifications.
- Adopt managed evolution to steer CPSoS evolution in a way such that necessary changes are implemented and the CPSoS remains flexible concerning future changes.
- Implement an SoS authority that has the capabilities to change RUI specifications, to assess the evolutionary state of the CPSoS, and to give incentives to control the onset of the evolutionary process.
- Define evolutionary steps that move the CPSoS towards its changed goal, but limit them in scope such that they remain predictable (with respect to the current knowledge of the CPSoS consortium) in their effects.
- Use the global timebase to temporally coordinate evolutionary steps.
- Use executable CPSoS models (see Sect. 4.2) in simulations and historic data recorded in the evolving CPSoS during runtime to pre-validate planned evolutionary steps.
- Use monitoring and assertion checks at the RUIs (e.g., by management facilities of the CPSoS dynamicity) to validate evolutionary steps. In case of hints about the onset of unexpected emergence update the models and take corrective actions, if possible before a detrimental effect manifests.
- Unexpected emergence appears to be unpredictable in principle even in carefully managed evolution. First occurrence might not be preventable in all cases, therefore use the mitigation strategy described in Sect. 5.4.

- Keep human domain experts in the design loop of evolutionary steps. The CPSoSs that we want to engineer and operate integrate humans, affect humans and should co-evolve with humans. Consequently, only humans are fully capable of judging how to best address a change in the environment of CPSoSs.

6 Conclusion

In this chapter we discussed interfaces in time-aware Cyber-Physical Systems-of-Systems (CPSoSs) for the purpose to investigate behavioral properties of CPSoSs. First, we characterized relevant architectural elements of CPSoSs and introduced three interface layers: the cyber-physical layer, the informational layer, and the service layer. Based on this conceptual groundwork, we identified (among other interfaces) the Relied Upon Interface (RUI) of a Constituent System (CS) as the fundamental interface responsible for the operational behavior of CPSoSs and managing CPSoS evolution.

The RUI is a CS interface on which the global, operational CPSoS service relies upon. We described the RUI model at each of the introduced interface layers, and outlined execution semantics for the informational layer to support the exploration of behavioral effects, like the occurrence of emergence.

In the second part of the chapter we focused on CPSoS evolution and how to manage it at the RUI of CSs. To this end we introduced an SoS authority that is in control of RUI specifications, plans evolutionary steps, and carries them out by changing RUI specifications. We discovered that both – changes in the CPSoS environment and evolutionary changes – harbors the risk of unpredicted detrimental emergence and suggested a mitigation strategy.

Acknowledgments. Warm regards to Sorin Iacob and Andrea Bondavalli for the insightful discussions about stigmergic channels. We thank the following experts for reviewing and helping to improve the chapter: Wilfried Elmenreich (Alpen-Adria-Universität Klagenfurt, AT), Sorin Iacob (Thales, NL), and Wilfried Steiner (TTTech, AT).

References

1. Air Accidents Investigation Branch: Aircraft Accident Report AAR1/2014 – Boeing 777-236ER, G-YMMM, 17 January 2008. Formal report (2008). https://www.gov.uk/aaib-reports/1-2010-boeing-777-236er-g-ymmm-17-january-2008. Accessed 1 Sept 2016
2. Al-Fuqaha, A., et al.: Internet of things: a survey on enabling technologies, protocols, and applications. IEEE Commun. Surv. Tutorials **17**(4), 2347–2376 (2015)
3. Alonso, G., et al.: Web Services. Data-Centric Systems and Applications, pp. 123–149. Springer, Heidelberg (2004)
4. Avižienis, A., Laprie, J.-C., Randell, B.: Dependability and Its Threats: A Taxonomy. Building the Information Society, pp. 91–120. Springer, Boston (2004)
5. Banks, A., Gupta, R.: MQTT Version 3.1.1. OASIS standard (2014)

6. Conrado, C., de Oude, P.: Scenario-based reasoning and probabilistic models for decision support. In: 2014 17th International Conference on Information Fusion (FUSION). IEEE (2014)
7. Caffall, D.S., Michael, J.B.: Architectural framework for a system-of-systems. In: 2005 IEEE International Conference on Systems, Man and Cybernetics, vol. 2. IEEE (2005)
8. Curbera, F., et al.: Unraveling the web services web: an introduction to SOAP, WSDL, and UDDI. IEEE Internet Comput. **6**(2), 86 (2002)
9. Da Xu, L., He, W., Li, S.: Internet of things in industries: a survey. IEEE Trans. Industr. Inf. **10**(4), 2233–2243 (2014)
10. Elmenreich, W., Haidinger, W., Kopetz, H.: Interface design for smart transducers. In: Proceedings of the 18th IEEE Instrumentation and Measurement Technology Conference, IMTC 2001, vol. 3. IEEE (2001)
11. Engel, J., Sturm, J., Cremers, D.: Camera-based navigation of a low-cost quadrocopter. In: 2012 IEEE/RSJ International Conference on Intelligent Robots and Systems. IEEE (2012)
12. Erl, T.: Service-Oriented Architecture: A Field Guide to Integrating XML and Web Services. Prentice Hall PTR, Upper Saddle River (2004)
13. Eugster, P.T., et al.: The many faces of publish/subscribe. ACM Comput. Surv. (CSUR) **35** (2), 114–131 (2003)
14. Fan, Z., et al.: Smart grid communications: overview of research challenges, solutions, and standardization activities. IEEE Commun. Surv. Tutorials **15**(1), 21–38 (2013)
15. Fielding, R.T.: Architectural Styles and the Design of Network-based Software Architectures. Dissertation University of California, Irvine (2000)
16. Fisher, D.: An emergent perspective on interoperation in systems of systems (2006)
17. Fuggetta, A., Picco, G.P., Vigna, G.: Understanding code mobility. IEEE Trans. Soft. Eng. **24**(5), 342–361 (1998)
18. Halbwachs, N., et al.: The synchronous data flow programming language LUSTRE. Proc. IEEE **79**(9), 1305–1320 (1991)
19. Han, J., Pei, J., Kamber, M.: Data Mining: Concepts and Techniques. Elsevier, Amsterdam (2011)
20. Henzinger, T.A., Horowitz, B., Kirsch, C.M.: Giotto: a time-triggered language for embedded programming. In: Henzinger, T.A., Kirsch, C.M. (eds.) EMSOFT 2001. LNCS, vol. 2211, pp. 166–184. Springer, Heidelberg (2001). doi:10.1007/3-540-45449-7_12
21. Höftberger, O., Obermaisser, R.: Ontology-based runtime reconfiguration of distributed embedded real-time systems. In: 16th IEEE International Symposium on Object/Component/Service-Oriented Real-time Distributed Computing (ISORC). IEEE (2013)
22. Khaleghi, B., et al.: Multisensor data fusion: a review of the state-of-the-art. Inf. Fusion **14** (1), 28–44 (2013)
23. Kohonen, T.: The Basic SOM. Self-organizing Maps, pp. 105–176. Springer, Heidelberg (2001)
24. Kopetz, H., Suri, N.: Compositional design of RT systems: a conceptual basis for specification of linking interfaces. In: Sixth IEEE International Symposium on Object-Oriented Real-Time Distributed Computing, 2003. IEEE (2003)
25. Kopetz, H.: Real-time Systems: Design Principles for Distributed Embedded Applications. Springer Science & Business Media, Heidelberg (2011)
26. Kopetz, H.: Why a Global Time is Needed in a Dependable SoS. arXiv preprint arXiv:1404.6772 (2014)
27. Kopetz, H.: A conceptual model for the information transfer in systems-of-systems. In: 2014 IEEE 17th International Symposium on Object/Component/Service-Oriented Real-Time Distributed Computing. IEEE (2014)

28. Kopetz, H., Frömel, B., Höftberger, O.: Direct versus stigmergic information flow in systems-of-systems. In: 2015 10th IEEE System of Systems Engineering Conference (SoSE) (2015)
29. Mahnke, W., Leitner, S.-H., Damm, M.: OPC Unified Architecture. Springer Science & Business Media, Heidelberg (2009)
30. Maier, M.W.: Architecting principles for systems-of-systems. INCOSE Int. Symp. **6**(1) (1996)
31. Mogul, J.C.: Emergent (mis) behavior vs. complex software systems. ACM SIGOPS Operating Syst. Rev. **40**(4), 293–304 (2006). ACM
32. Murer, S., Bonati, B., Furrer, F.J.: Managed evolution (2011)
33. Potop-Butucaru, D., de Simone, R., Talpin, J.-P.: The synchronous hypothesis and synchronous languages. In: The Embedded Systems Handbook, pp. 1–21 (2005)
34. Quigley, M., et al.: ROS: an open-source Robot Operating System. ICRA Workshop Open Source Softw. **3**(3.2), 5 (2009)
35. Steiner, W., et al.: Ttethernet dataflow concept. In: 2009 Eighth IEEE International Symposium on Network Computing and Applications, NCA 2009. IEEE (2009)
36. Valckenaers, P., Kollingbaum, M., Van Brussel, H.: Multi-agent coordination and control using stigmergy. Comput. Ind. **53**(1), 75–96 (2004)
37. El Salloum, C., et al.: The ACROSS MPSoC–a new generation of multi-core processors designed for safety–critical embedded systems. Microprocess. Microsyst. **37**(8), 1020–1032 (2013)

Emergence in Cyber-Physical Systems-of-Systems (CPSoSs)

Hermann Kopetz[1], Andrea Bondavalli[2], Francesco Brancati[3], Bernhard Frömel[1(✉)], Oliver Höftberger[1], and Sorin Iacob[4]

[1] Institute of Computer Engineering, Vienna University of Technology, Vienna, Austria
h.kopetz@gmail.com, {froemel, oliver}@vmars.tuwien.ac.at

[2] Department of Mathematics and Informatics, University of Florence, Florence, Italy
andrea.bondavalli@unifi.it

[3] Resiltech SRL, Pisa, Italy
francesco.brancati@resiltech.com

[4] Thales Netherlands B.V, Delft, The Netherlands
sorin.iacob@nl.thalesgroup.com

1 Introduction

The essence of the concept emergence is aptly communicated by the following quote, attributed to Aristotle, who lived more than 2000 years ago: *The Whole is Greater than the Sum of its Parts*. The interactions of *Parts* can generate a *Whole* with unprecedented properties that go beyond the properties of any of its constituent *Parts*. The immense varieties of inanimate and living entities that are found in our world are the result of emergent phenomena that have a small number of elementary particles at their base.

A *System-of-Systems (SoS)* consists of a set of autonomous technical systems, called *constituent systems (CS)* that are independent and provide a useful service to their environment [18]. The purpose of building a System-of-Systems out of CSs is to realize new services that go beyond the services provided by any of the isolated CSs. *Emergence is thus at the core of SoS engineering.*

A Cyber-Physical System (CPS) is a synthesis of processes in the physical environment and computer systems that contain sensors to observe the physical environment and actuators to influence the physical environment. In most cases, the computer systems are distributed and contain computational nodes connected through networks that realize the information exchange among the nodes. A Cyber-Physical System-of-Systems (CPSoS) is an integration of stand-alone CPSs that provides services that go beyond the services of any of its isolated CPSs.

It is the objective of this chapter to investigate the phenomenon of emergence in CPSoS. In the following section we look at some prior work on emergence in the

This work has been partially supported by the FP7-610535-AMADEOS project.

A. Bondavalli et al. (Eds.): Cyber-Physical Systems of Systems, LNCS 10099, pp. 73–96, 2016.
DOI: 10.1007/978-3-319-47590-5_3

domains of philosophy and computer science. Since emergence is always referring to phenomena that occur at a given level of a hierarchic system model, Sect. 3 elaborates in detail on the concept of a *multi-level hierarchy*. Section 4 presents a definition of emergence in the SoS context and discusses some properties of emergent phenomena. Section 5 introduces a number of examples of emergent phenomena in computer systems. Section 6 discusses some design guidelines that help to detect the potential of emergent phenomena in a CPSoS and mitigate the effects of detrimental emergence. This Chapter terminates with a conclusion in Sect. 7.

2 Related Work

In philosophy the questions of how the diversity of the world emerges out of simple physical building blocks has been a topic of inquiry since the time of the ancient Greeks, leading to abundant literature about emergence, e.g., the survey articles [20, 34] or the books by [4, 6, 16]. Computer scientists got interested in the topic of emergence when it was realized that some striking phenomena that are observed at the system level of complex systems could not be explained by looking at the system's components in isolation. A well-publicized example of such a striking phenomenon is the flash crash of the stock market on May 6, 2010 [2]. Emergence can be regarded as an intriguing *part-whole relation* that investigates how the properties and the interaction of the parts lead to novel phenomena of a whole.

Holland remarks in [16]: *Despite its ubiquity and importance, emergence is an enigmatic and recondite topic, more wondered at than analyzed... It is unlikely that a topic as complicated as emergence will submit meekly to a concise definition and I have no such definition to offer*. Fromm [9, 10] elaborates on different forms of emergence and investigates the emergence of complexity in large systems. In [26], Mogul describes emergent misbehavior in a number of computer systems, discusses how emergence can manifest itself, and proposes a research agenda for studying the phenomena of emergence in complex computer systems. In the European Research Project TAREA SoS the current state of the art in the field of SoS has been captured [14] and a roadmap for future SoS research has been proposed. In this roadmap the topics of theoretical foundations of SoSs and of emergence are in a prominent position. In [19], Keating argues for the development of a firm epistemological foundation of emergence in SoSs. In the proceedings of the yearly IEEE conference on Systems of Systems Engineering and the book [18] by Jamshidi relevant contributions to the topic of emergence in SoSs can be found. Parunak and VanderBrok [28] and Huberman and Hogg [17] observed that variable temporal delays play a key role in the generation of emergent misbehavior in an SoS. In [5] Boschetti and Gray elaborate on the limits of insights gained from computer simulations when modeling emergent phenomena in natural systems.

3 Multi-level Hierarchy

The understanding and analysis of the immense variety of things and their behavior in the non-living and living world around us require appropriate modeling structures. Such a modeling structure must limit the overall complexity of a single model and support the step-wise integration of a multitude of different models. One such widely identified modeling structure is that of a *multi-level hierarchy*, where level-specific rules and laws govern the interdependence of entities at each level of the hierarchy. Since the phenomenon of emergence is always associated with levels of a *multi-level hierarchy* it is useful to start with a thorough discussion of *multi-level hierarchies.*

A *multi-level hierarchy* is a recursive structure where a system, the *whole* at the level of interest (the *macro-level*), can be *taken apart* into a set of sub-systems, the *parts*, that *interact* statically or dynamically at the level below (the *micro-level*). Each one of these sub-systems can be viewed as a system of its own when the focus of observation is shifted from the level above to the level below. This recursive decomposition ends when the internals of a sub-system is of no further interest. We call such a sub-system at the lowest level of interest (the *base* of the hierarchy) an *elementary part* or a *component.*

In his seminal paper The Architecture of Complexity Herbert Simon posits [32] (p. 219): If there are important systems in the world that are complex without being hierarchic, they may to a considerable degree escape our observation or understanding.

Our models of the *world of things* are organized along such a widely cited *Multi–level Material Hierarchy*, giving rise to the establishment of dedicated scientific disciplines for each level, e.g.:

- Atoms consist of elementary particles (the field of physics)
- Molecules consist of atoms (the field of chemistry)
- Cells consist of molecules (the field of biology)
- Organs consist of cells (the field of medicine).

3.1 Whole versus Parts

Viewed from the macro-level, the whole is an *established entity* that encapsulates and hides its parts that interact at the lower level. If the parts at the micro-level that form the whole at the macro level are all identical we talk about a *homogeneous* structure, otherwise we talk about a *heterogeneous* structure.

At a given macro-level, we consider the whole as an entity that is surrounded by a *surface*. Interfaces located at the surface of the whole control the exchange of *matter*, *energy* or *information* among the wholes at the same level.

Koestler [21] (p. 341) has introduced the term *holon* to refer to the *two-faced character* of an entity in a multi-level hierarchy. The word *holon* is a combination of the Greek "holos", meaning *all*, and the suffix "on" which means *part*. The point of view of the observer determines which view of a given holon is appropriate in a particular scenario.

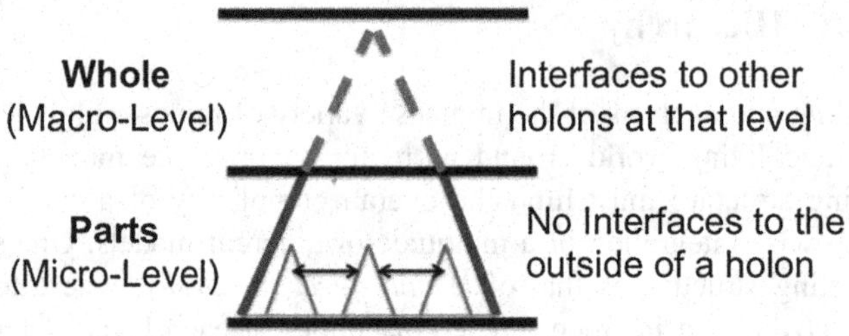

Fig. 1. Two-faced character of a holon

Figure 1 gives a graphical representation of the holon. Viewed from the outside at the macro level, a holon is a stable whole that can interact with other holons of that level by an interface across its surface. Viewed from below, the micro-level, a holon is characterized by a set of interacting parts that are confined by the boundaries of the holon. **This rigorous enclosure of the parts of a holon at the micro-level is absolutely essential to maintain the integrity of the abstraction of a holon as a whole at the macro level.**

Koester states in [21] (p. 343): Every holon has the dual tendency to preserve and assert its individuality as a quasi-autonomous whole; and to function as an integrated part of an (existing or evolving) larger whole. This polarity between Self-Assertive (S-A) and Integrative (INT) tendencies is inherent in the concept of hierarchic order and a universal characteristic of life.

There are two relations characterizing two adjacent levels of a hierarchy: (i) the ***level*** *relation* between the *whole* at the macro-level and the *parts* of the micro-level and (ii) the ***interaction*** *relation* among the *parts* of the micro-level.

3.2 Level Relations

The type of the *level relation* determines the character of a multi-level hierarchy. In this section we focus on three types of level relations, a *nested (or structure) hierarchy*, a *description hierarchy* and a *control hierarchy*. For the emergence of novel behavior in a CPSoS the control hierarchy is the most important.

Structure Hierarchy. We call a hierarchy a *structure (or nested) hierarchy* if the *whole comprises the parts* or, in different wording, the *parts are contained in the whole*, i.e., *consists of* (from the top to the bottom) or *forms* (from the bottom to the top) stand for the *level relation* of *containment.*

Structure hierarchies are formed by the identification and classification of the observation of physical structures that are existent in the world of things, irrespective of the subjective view of the observer. These physical structures are often formed by *physical force-fields* (see also Sect. 3.3, Physical Interactions).

The *Multi-level Material Hierarchy* referred to in the beginning of Sect. 3 above is an example for a *structure hierarchy*.

Description Hierarchy. A multi-level hierarchy that describes a *set of related entities* at different levels of abstraction is called a *multi-level description hierarchy*. A description hierarchy can be much simpler than the related *structure hierarchy* provided the structure hierarchy is highly redundant. If a complex structure is completely un-redundant, then it is its own simplest description [32] (p. 221).

We distinguish two types of descriptions, *state descriptions* and *process descriptions*. State descriptions describe the state of the world at the *instant of observation*. *Process descriptions* explain how a new state of the world unfolds as time progresses that is how the state transitions happen. *A description of behavior* is a process description.

The classification of entities in a description hierarchy is usually based on cognitive models of the observer and thus may be dependent on the subjective view of the observer. Moreover, depending on the purpose, different levels of description of the same physical structure can be introduced by the observer.

For example, the *thermodynamic description* of the behavior of a gas is at a higher level of description than the *statistical description* of the *same physical material* and the choice among them may depend on the purpose of the description.

If the redundancy of a structure is removed from its description hierarchy, then a significant simplification of the description can be realized (e.g., [32] p. 220).

In case the elements of a hierarchy are *constructs*, i.e. non-material entities that are the product of the human mind, the assignment of the constructs to hierarchical levels always results in a description hierarchy, the organization of which is determined by the purpose of the observer.

In many, but not in all cases, the description hierarchy of a structure follows the structure hierarchy.

Control Hierarchy. In a control hierarchy the macro-level provides some *constraints* on the structure or behavior of the parts at the micro-level thus establishing a causal link from the macro level to the micro-level. Constraints restrict the behavior of things beyond the natural laws, which the things must always obey.

In many, but not all cases, the *control hierarchy* follows the *structure hierarchy*. Ahl [1] (p. 107) provides the following example: The concept *army* denotes a structure hierarchy that *consists of the soldiers of all ranks and contains them all.* In contrast, *a general* at the top of an army (a military hierarchy) *controls* the soldiers, but *does not contain* them.

In some cases, as the example of the military hierarchy above shows, the control constraints originate from *outside*, i.e. above the macro-level. In other cases, the control constraints have their origin in the *whole*, i.e. the *collective behavior* of the parts of the micro-level. It is this latter case that is relevant for the analysis of emergence. Many equivalent examples can be found in Distributed Computing when we have centralized or decentralized control and management. Since behavior (function plus time) is a concept that depends on the progression of time, there is a temporal dimension in control hierarchies that deal with behavior.

Since the behavior of the parts forms the behavior of the whole, but the whole can constrain the behavior of the parts we have an example of a *causal loop* in such a control hierarchy.

We can observe such a causal loop in many scenarios that are classified as *emergent* in every-day language: the behavior of birds in flocks, the synchronized oscillations of fireflies or the build-up of a traffic jam at a congested highway.

Pattee [30] discusses control hierarchies extensively in *The Physical Basis and the Origins of Hierarchical Control.* In order to support the simplification at the macro-level and establish a hierarchical control level, a control hierarchy must on one side *abstract from* some degrees of freedom of the behavior of the parts at the micro-level but on the other side must *constrain* some other degrees of freedom of the behavior of the parts, i.e., a control hierarchy must provide *constraints from above*, while, in a multi-level material hierarchy the natural laws provide *constraints from below*.

The delicate borderline between *the constraints from above on the behavior of the micro-parts* and *the freedom of behavior of the micro-parts* is decisive for the proper functioning of any control hierarchy.

There are two extremes of control which lead to a collapse of the control hierarchy: (i) full control from above which defeats the principle of abstraction of control and leads to a full deterministic behavior and (ii) no constraints from above which can lead to unconstrained chaotic behavior (see Fig. 2).

For example, a good conductor of an orchestra will control the tempo of the performance without taking away the freedom from the musicians to express their individual interpretation of the music.

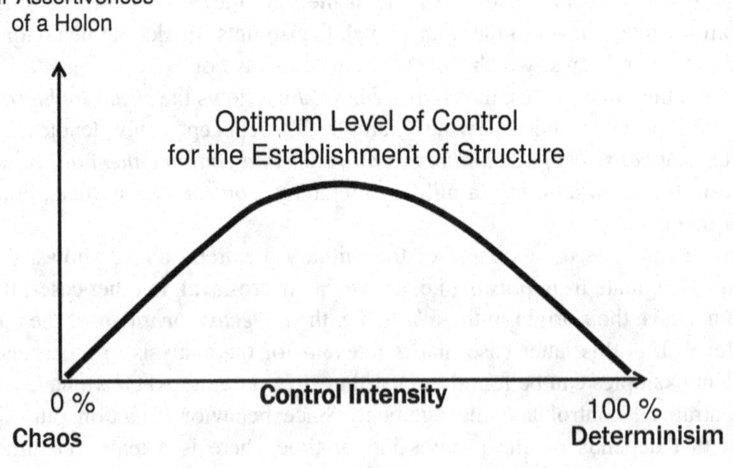

Fig. 2. Self assertiveness of a holon

3.3 Interaction Relations

Formal Hierarchy. Simon [32] (p. 195) calls a hierarchy a *formal hierarchy* if the *interaction relation* is empty, i.e., the parts are only related to the whole of the higher adjacent level. If, in the above example, the soldiers relate at a given level only to their boss, but not to each other, then we have an example of a formal hierarchy. Models that have the structure of a formal hierarchy are rare.

Physical Interactions. The *physical interactions* at any considered level of a material hierarchy can be classified in the following three dimensions: (i) *distance* among the parts, (ii) *force fields* among the parts and (iii) *frequency of interactions* among the parts. In general, as we move up the levels of a material hierarchy the *distance increases*, the *force-field magnitude decreases* and the *frequency of interactions* decreases [32].

Simon argues that the laws that govern the behavior at each level are *nearly independent* of the level above and below, giving rise to the principle of *near decomposability* [32] (p. 209) of levels.

This principle of *near decomposability* states that an approximate model suffices in most cases to model the behavior at any given level of a multi-level hierarchy.

This approximate model considers only the physical interactions at the considered level and abstracts from the behavior of the *high-frequency parts at the level below* and considers the dynamic parameters of the *low frequency parts at the level above* that provide the constraints as *constants*.

Informational Interactions. Informational interactions exchange information among the communicating partners. When the information exchanged consists of *data* and an *explanation of the data* we observe the exchange of *Itoms*.

Itom: An Itom is an atomic unit of *object data* and *meta data*. The object data represents some semantic content, and the meta data provides an explanation of the object data, i.e., how the semantic content represented by object data can be accessed. The semantic content of (or the information contained in) an Itom reports about a timed proposition relating to some entities in the world [23].

In a Cyber-Physical System-of-Systems (CPSoS) we distinguish between two types of informational interactions: (i) *message-based information* interactions in cyber space and (ii) *stigmergic information* interactions in the physical world.

Interactions in the cyber space allow in principle the exchange of explicitly defined Itoms which travel unmodified (invariant semantic content) from a sender to a set of receivers. Stigmergic interactions are indirect and involve influencing the state of the common environment of senders and receivers. Such environment may also be under the possible influence of *environmental dynamics*. Environmental dynamics are autonomous processes in the environment (physical world or cyber space) that also act on the state of the environment. Consequently, in stigmergic interactions it is – in many cases – not possible to send the same Itom from sender to receivers. Instead very often receivers will only be able to observe object data which is (more or less closely) related to the original data sent and needs to be correctly interpreted to avoid *property mismatch*. A model of the environmental dynamics able to represent the processing and modifications performed on data would be paramount in the understanding and mastering of stigmergic information exchange.

In cyber space data is represented by a bit-pattern that can be generated by the processing of stored Itoms or by some data acquisition process, e.g., by a sensor. For data acquisition, the design of the sensor determines how the acquired bit pattern has to be interpreted, i.e., provides for the explanation of the object data.

Since an *Itom* is a higher-level concept than the sole *object data* in an Itom, we propose to use *Itoms* in the specification of *Relied-Upon Interfaces (RUIs)* among the Constituent Systems (CSs) of a CPSoS (see Chap. 2). According to [23] the full specification of an Itom has to provide answers to the following questions:

- **Identification**: *What entity is involved?* The entity must be clearly identified in the space-time reference frame.
- **Purpose**: *Why is the data created?* This answer establishes the link between the raw data, the refined data and the purpose of the CPSoS.
- **Meaning**: *How has the data to be interpreted by a human or manipulated by a machine?* If the answer to this question is directed towards a human, then the presentation of the answer must use symbols and refer to concepts that are familiar to the human. If a computer acquires data, then the explanation must specify how the data must be manipulated and stored by the computer.
- **Time**: *What are the temporal properties of the data?* Real-time data must include the instant of observation in the entity. In control applications it is helpful to include a second timestamp, a *validity instant* that delimits the validity of the control data as part of the Itom [22] (p. 4).

Message-based Information Flows: A *message-based information flow* is present if one CS sends a message to another CS. In many legacy distributed systems only *object data* is contained in a message while the *explanation of the data* is derived from the context.

In a CPSoS the involved CSs can be operating in differing contexts, e.g., in the *US* and *Europe*. For example, in the US temperature is represented by *degrees Fahrenheit*, while in Europe temperature is represented by *degrees Celsius*. As a consequence, the *same data (bit-patterns)* can convey a different meaning if the contexts of the sender differs from the context of the receiver of the message, causing a *property mismatch*. Such *property mismatches* have been the cause of severe accidents.

Stigmergic Information Flows: A *stigmergic information flow* is present if one sending CS acts on the physical environment and changes the state of the environment and later on another receiving CS observes the changed state in the environment with a sensor that captures the *sensor specific aspect* of the environment [24]. Consider, for example, the coordination of cars on a busy highway to realize a smooth flow of traffic. In addition to the direct communication by explicit signals among the drivers of the cars (e.g., the blinker or horn), the *stigmergic information flow* based on the observation of the movement of the vehicles on the road (caused by the actions of other drivers) is a primary source of information for the assessment of a traffic scenario. An important characteristic of stigmergic information flows is the consideration of up to date *environmental dynamics*.

Hidden Channels. There exist many indirect information flows, in particular stigmergic ones, which remain both (i) unknown to the sender which is not aware of the flow, and (ii) are not captured by systems designers or modelers. We call such existing interaction relations *hidden channels*.

Hidden channels are problematic, because they can contribute to the generation of causal loops (and therefore take active part in the rise of emergent phenomena). In addition, these causal links may lead to **a modification of the understood holarchy abstraction,** i.e., parts of one level interact directly with parts of another level **which may establish hidden level relations** (e.g., a control hierarchy). Effects of such modification of the holarchy abstraction may cause both unintended information leakage (violations of security properties) and unexpected negative emergence.

Usually it is difficult to protect the state of the physical environment regarding observations of receivers. Additionally, in many cases a sender may be even unaware of leaking information to its environment. For example, consider security attacks based on observing the electromagnetic emissions of a processor on smart cards [11].

Still, hidden channels should be avoided by properly identifying them (see Sect. 6.1) or insulating against them (e.g., firewalls, physical insulation).

4 Emergence

It is quite common, as we move up a multi-level hierarchy, that novel phenomena can be observed at a given level that are not present at the level below. We call these new phenomena *emergent phenomena*. We use the term *phenomenon* as an umbrella term that can refer to *structure*, *behavior* or *property*.

In many cases the laws that explain the genesis of these emergent phenomena are formulated *post facto* because it would require a *very knowledgeable mind* to predict *a priori* all possible phenomena that can come into existence out of the interactions of many given parts. The first appearance of an emergent phenomenon is *often a surprise* to a human observer.

4.1 Definition of Emergence

In order to achieve a level of objectivity we aim for a definition of emergence that is based on a *property of the scenario* and not on a *relation* between the scenario and the observer.

Let us analyze the relationship between two adjacent levels of a multi-level hierarchy, the *micro-level* (the level of the *parts*) and the *macro-level* (the level of the *whole*) where emergent phenomena are observed, assuming that the *level relation is given*. We restrict our analysis to these two levels and disregard the case where some properties of the parts are themselves emergent with respect to their lower-level parts. Our definition of emergence in a Cyber-Physical Systems-of-Systems is the result of many interdisciplinary discussions during the *AMADEOS Workshop on Emergence in Cyber-Physical Systems-of-Systems* [15].

A phenomenon of a whole at the macro-level is emergent if and only if it is *of a new kind* with respect to the non-relational phenomena of any of its proper parts at the micro level.

A phenomenon is of *a new kind* if the concepts required to *explain* this phenomenon cannot be found in the world of the isolated parts. *Conceptual Novelty* is thus the landmark of our definition of emergence.

Note that, according to the above definition, the emergent phenomena *must only be of a new kind with respect to the non-relational phenomena of the parts*, not with respect to the knowledge of the observer. If a phenomenon of a whole at the macro-level is *not of a new kind* with respect to the non-relational phenomena of any of its proper parts at the micro level then we call this phenomenon *resultant*.

The essence for the occurrence of emergent phenomena at the macro-level (the *SoS level*) lies in the *interactions of the parts* at the micro-level, i.e., in the spatial arrangement of the parts caused by physical force-fields and/or the *designed temporal informational interactions* among the parts at the micro-level.

In CPSoS, the phenomenon we are interested in is *behavior*. In a CPSoS the *observable behavior* of a system is *the temporal sequence of observable states* of the system in the *Interval of Discourse*. We are thus interested in *diachronic* emergence, where initial interactions of the parts at the micro-level precede the appearance of the emergent phenomenon at the macro level.

We assume that the temporal distance between two observation instants of an observer is a multiple of a smallest duration. This smallest temporal distance expresses the *grain of observation* of this particular observer. If the duration of a state is shorter than the *grain of observation* then this short-lived state may evade the observations of this observer. The duration of the grain of observation should be selected on the basis of the *purpose of the observer*, the *dynamics of the observed system* and the *minimal response time* of the entities at the chosen level of observation.

Some scientists posit that emergent behavior is connected with a *surprise of the observer* [31]. According to this view, emergence occurs, if the causal link between the *interactions of the parts* and the *behavior of the whole* is *non obvious* to the observer (and therefore a surprise to the observer). According to this definition, the *state of knowledge of the observer* is the decisive criterion for the classification of a phenomenon as *emergent*. As a consequence, different observers with different states of knowledge will judge the same phenomenon differently. It follows that emergence is considered a *relation* between the whole and the *observer* and not a *property* of the whole.

4.2 Explained vs Unexplained Emergence

At first we pose the question whether emergent properties are *reducible* to the properties of the parts considered in isolation.

The following quote about Scientific Reduction is taken from the Stanford Encyclopedia on Philosophy:

The term 'reduction' as used in philosophy expresses the idea that if an entity x reduces to an entity y then y is in a sense prior to x, is more basic than x, is such that x fully depends upon it or is constituted by it. ***Saying that x reduces to y typically implies that x is nothing more than y or nothing over and above y.***

In an *artifact*, such as a CPSoS, emergent properties appear at the *macro-level* if the parts at the *micro-level* interact according to a *design provided by a human designer*—this is ***more*** than the parts considered in isolation. It follows that *emergent properties* in a CPSoS are not *reducible* to the parts considered in isolation.

According to our definition of emergence in Sect. 4.1, a novel phenomenon is considered *emergent*, irrespective of whether it can be *explained* how the new phenomenon at the macro level has developed out of the parts at the micro-level. Given the *present state of knowledge*, some of these emergent phenomena can be explained by existing theories while there are other emergent phenomena where at present no full explanation can be given as to how they developed. Examples for (as of today) unexplained emergence are the *generation of life* or the *generation of the mind* on top of the neurons in the brain.

But what constitutes a *proper scientific explanation*? Hempel and Oppenheim [13] (p. 138) outlined a general schema for a *scientific explanation* of a phenomenon as follows:

> **Given**
> *Statements of antecedent conditions*
> **and**
> *General Laws*
> **then a logical deduction of the**
> *Description of the empirical phenomenon to be explained*
> **is entailed.**

The *antecedent conditions* can be initial conditions or boundary conditions that are *unconstrained* by the general laws.

The *general laws* can be either universally valid *natural laws* that reign over the behavior of things or *logical laws* describing a valid judgment in the domain of constructs. Natural laws do not change in time or have a memory of the past. A natural law, such as a physical law, must hold everywhere, no matter what level of a multi-level hierarchy is the focus of the investigations.

A weaker form of explanation is provided if the *general laws* in the above schema are replaced by *established rules.* There are fundamental differences between general laws and established rules. General laws are *inexorable* and *universally* valid while established rules are *structure dependent* and *local*. Rules about the behavior of things are based on more or less meticulous experimental observations. A special case is the introduction of *imposed rules*, e.g., the rules of an artificial game, such as chess. The degree of accuracy and rigor of various established rules differ substantially.

It thus follows that between the two extremes of *scientifically explained* and *not explained at all* there is a *continuum of explanations* that are more or less acceptable and are relative with respect to the general state of knowledge and the opinion of the observer at a given point in time.

4.3 Conceptualization at the Macro-level

According to our definition of emergence, novel concepts should be formed and new laws may have to be introduced to be able to express the emerging phenomena at the macro level appropriately. Note that the emergent phenomena and laws must be new w. r.t. the phenomena of the isolated parts, but not necessarily new with respect to the knowledge of the observer, i.e., such phenomena are emergent irrespective of the state of knowledge of the observer.

In the history of science, many novel laws that employ *new concepts* have been introduced to capture the newly observed regularities of phenomena at a macro-level. We call such a new law that deals with the emerging phenomena at a macro level *an intra-ordinal law* [27]. At a later time, some of these laws have been reduced to well-understood effects of the parts at the adjacent micro-level, e.g., the *thermodynamic theory of a gas* can be explained by the *statistical theory of gas* [3].

Since the concepts at the macro level are new with respect to the existing concepts that describe the properties of the parts, the established laws that determine the behavior of the parts at the micro-level will probably not embrace the new concepts of the macro-level. Therefore, it is often necessary to formulate *inter-ordinal laws* (also called *bridge laws*) to relate the established concepts at the micro-level with the new concepts of the macro-level.

The proper conceptualization of the new phenomena at the macro level is at the core of the simplifying power of a multi-level hierarchy with emergent phenomena.

Let us look at the example of a transistor. The *transistor effect* is an emergent effect caused by the proper arrangement of dopant atoms in a semiconducting crystal. The exact arrangement of the dopant atoms is of no significance as long as the provided behavioral specifications of a transistor are met. In a VLSI chip that contains millions of transistor, the detailed microstructure of every single transistor is probably unique, but the external behavior of the transistors (the holons) is considered the *same* if the behavioral parameters are within the given specifications. It is a tremendous simplification for the designer of an electronic circuit that she/he does not have to consider the unique microstructure of every single transistor.

4.4 Downward Causation

In classical physics, the concept of causation links an *effect* to an earlier *cause*. If in the domain of Newtonian mechanics precisely defined initial conditions (the *cause*) are given, an object will move along a trajectory (the *effect*) that is fully determined by the differential equations that express the laws of macro-mechanics. However, in the domain of micro-mechanics, where quantum-physical laws reign, it is not possible to observe the initial conditions of an object without influencing the object of observation. This is one of the reasons, why the concept of unidirectional causation is highly debated in the modern sciences. Another reason pertains to the multitude of parameters, captured in the notion of a *causal field* that characterizes the causes of real-life phenomena. It is often up to subjective judgment to determine which one of these many causes is considered the *most prominent cause.*

On the other side, the *unidirectional cause-effect* relation plays a prominent role in our subjective models of the world in order to realize intended effects or to avoid the *causes* of *undesired effects*. To quote Pattee [29] (p. 64 onwards): *I believe the common everyday meaning of the concept of causation is entirely pragmatic. In other words, we use the word cause for events that might be controllable... the value of the concept of causation lies in its identification of where our power and control can be effective. ... when we seek the cause of an accident, we are looking for those particular focal events over which we might have had some control. We are not interested in all those parallel subsidiary conditions that were also necessary for the accident to occur, but that we could not control... .*

Along this line of reasoning the *term downward causation* denotes the concept that the whole at the macro-level can *constrain* or even *control* the behavior of the parts at the micro-level (the level below).

Downward causation is a difficult concept to define precisely, because it describes the collective, concurrent, distributed behavior at the system level. ... Downward causation is ubiquitous and occurs continuously at all levels, but it is usually ignored simply because it is not under our control. ... The motion of one body in an n-body model might be seen as a case of downward causation [29] (p. 64).

Downward causation establishes a *causal loop* between the *micro-level* and the adjacent *macro level*. The interaction of the parts at the micro-level causes the whole at the macro-level while the whole at the macro-level constrains the behavior of the parts at the micro-level (see also Sect. 5.2). We conjecture that in a multi-level hierarchy emergent phenomena are likely to appear at the macro-level when there is a causal-loop formed between the micro-level that forms the whole and the whole (i.e., the ensemble of parts) that constrains the behavior of the parts at the micro-level.

In a system that exhibits downward causation the degrees of freedom of the parts that can be exploited at the micro-level, e.g., by mechanisms of self-organization are limited by:

1. Constraints on the degrees of freedom of material parts at a micro-level coming from below, i.e., *upward causation* deriving from applicable *natural laws, e.g., the laws of physics*.
2. Constraints on the degrees of freedom of a part at the micro-level coming from above, the *whole* at the macro-level by *downward causation.*

Note that in a concrete system, some of these categories can be empty. For example, in a hierarchy of *constructs* there is no *upward causation*, i.e. constraints on the parts from below caused by natural laws.

In our opinion the *exclusion argument* by Kim [20] —that in a system with downward causation *macro causal powers compete with micro causal powers* and, if this is the case, micro causal powers will always win, needs to be reconsidered since the macro causal powers and the micro causal powers restrict different degrees of freedom of the parts and are thus not in conflict.

Another different way in which emergence is observed in practice in the real world also is the one caused by a *Cascade effect* [8]. A cascade effect exists, if in a system with a multitude of parts at the micro level a state change of a part at the micro-level causes successive state changes of many other parts at the micro level. The cumulative effect of the totality of these state changes results in a novel phenomenon, such as an *avalanche* or a *nuclear explosion*. An *epidemic* is also a good example for a cascade effect. Cascade effects are *diachronic*, since they develop over time.

There may be other mechanisms that lead to emergent phenomena that we have not yet identified.

4.5 Supervenience

The principle of *Supervenience* [25] establishes an important dependence relation between the emerging phenomena at the macro-level and the interactions and arrangement of the parts at the micro-level. *Supervenience* states that

Sup_1: *a given emerging phenomenon at the macro level can emerge out of many different arrangements or interactions of the parts at the micro-level while*

Sup_2: *a difference in the emerging phenomena at the macro level requires a difference in the arrangements or the interactions of the parts at the micro level.*

Because of *Sup_1* one can abstract from the many different arrangements or interactions of the parts at the micro level that lead to the same emerging phenomena at the macro level—see the example of the *transistor* above. *Sup_1* entails a *significant simplification* of the higher-level models of a multi-level hierarchy.

Because of *Sup_2* any difference in the emerging phenomena at the macro level can be traced to some significant difference at the micro level. *Sup_2* is important from the point of view of *failure diagnosis*.

4.6 Classification of Emergence

Figure 3 depicts a schema for the classification of emergent phenomena.

In a CPSoS the CSs interact, i.e., via message-based channels in cyber space in which they exchange *Itoms,* and interact also via stigmergic channels information flows in the

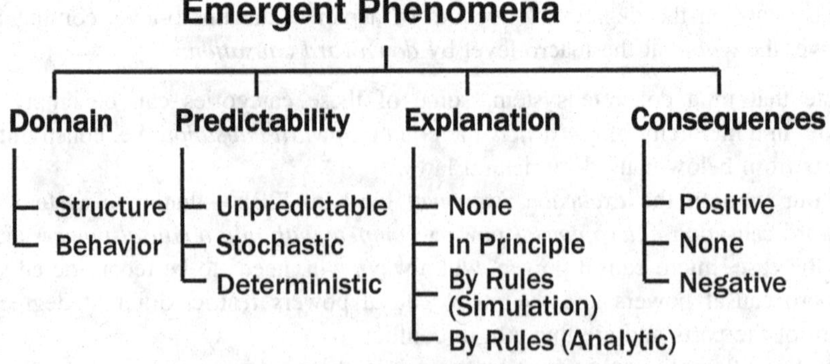

Fig. 3. Classification of emergent phenomena

physical world. These interactions can give rise to *emergent behavior* at the level of CPSoS. Although this behavior is explainable in principle, we may not be able to explain or predict this behavior in practice due to our ignorance about the full scope of the CPSoS, the precise temporal interactions among the CS (see e.g. the deadlock example in Sect. 3.5) and hidden communication channels behind the interfaces of a CS.

5 Examples of Emergence in Computer Systems

In this Section we discuss a number of examples of emergent behavior in computer systems. The first four examples can be explained, while the fifth example, the *Flash Crash* of the stock market on May 6, 2010 [2], although *explainable in principle* has not been *explained in practice* up to today.

5.1 Deadlock in Computer Systems

In some publications, the occurrence of a *deadlock* in a computer system is called an *emergent phenomenon* [12]. With the advent of multi-programming computer systems, the following event has been occasionally observed: when executing a number of processes concurrently, the system comes to a *permanent halt*, although each process, executed in isolation executes flawlessly. At first, this phenomenon could not be explained and was considered a *surprise*. Later on (around the year 1970) a *full explanation* of this phenomenon, called *deadlock*, was given [7]. The following simple example of Fig. 4 explains the essence of the phenomenon deadlock.

Let us consider the execution of a seat reservation system (cf. Fig. 4) in an *ideal world*, where no failures of the computer hardware will ever occur. As long as only a

Process Type A	**Process Type B**
1 $S^{money} = 1, S^{seat} = 1$	1 $S^{money} = 1, S^{seat} = 1$
2 Client selects seat and provides credit card	2 Client selects seat and provides credit card
3 **Wait** (S^{money})	3 **Wait** (S^{seat})
4 Get *Money*	4 Get *Seat*
5 If *No-Money* Then **Signal** (S^{money}) Print *No Money* Goto 2	5 If *No-Seat* Then **Signal** (S^{seat}) Print *No Seat* Goto 2
6 **Wait** (S^{seat})	6 **Wait** (S^{money})
7 Get *Seat*	7 Get *Money*
8 If *No-Seat* Then *Return Money* **Signal** (S^{money}) **Signal** (S^{seat}) Print *No Seat* Goto 2	8 If *No-Money* Then *Return Seat* **Signal** (S^{money}) **Signal** (S^{seat}) Print *No Money* Goto 2
9 **Signal** (S^{money}) **Signal** (S^{seat})	9 **Signal** (S^{money}) **Signal** (S^{seat})
10 Print *Seat Ticket*	10 Print *Seat Ticket*
11 Goto 2	11 Goto 2

Fig. 4. Example of deadlock

finite number of reservation processes of Type A are executed concurrently, the system will operate flawlessly forever. The same will happen if only a finite number of reservation processes of Type B execute concurrently. However, if a finite number of processes of Type A and processes of Type B operate concurrently, the system will sometimes *stop forever (deadlock)*. *Stopping forever* is the novel phenomenon that is not happening if processes of Type A or processes of Type B operate in isolation.

In the program sketch of Fig. 4 there are two semaphore variables, S^{money} and S^{seat} initialized with the value *1*. Whenever a process executes a *Wait* operation on a semaphore variable, the process is only allowed to enter the following *Critical Section* if the value of the semaphore variable is positive at the start of execution of the *atomic operation Wait*. The atomic operation *Wait* tests the value of the designated semaphore variable. In case the test gives a positive value, it decreases the value of the semaphore variable by 1 and enters the Critical Section. Otherwise it waits until the value of the semaphore variable gets positive. The semaphore operation *Signal*, executed at the end of a *Critical Section*, increases the value of the designated semaphore variable by *1* and thus enables another waiting process to enter the Critical Section.

In Fig. 4, the semaphore S^{money} ensures that in the following *Critical Section*, dealing with the *money* only a single process is allowed to execute at an instant. Likewise, the semaphore variable S^{seat} ensures that in the following *Critical Section* dealing with the *seat allocation* only a single process is allowed to execute at a time. As long as processes of type A execute concurrently, the execution of *Wait*(S^{money}) is always followed by *Wait*(S^{Seat}).

However, if the executions of processes of *Type A* and *Type B* are interleaved, then it can happen that a process of Type A enters the Critical Section protected by S^{money} and, before the process of Type A executes the operation *Wait*(S^{Seat}) a process of Type B enters its critical Section protected by S^{seat}. From now on, a deadlock is unavoidable if the *money* and the *seat* are available, since both processes have to *wait forever* on the release of the respective following *Critical Section*.

The observed phenomenon of *deadlock* fulfills the requirement of an emergent phenomenon:

- The phenomenon *deadlock—halting forever*—is novel with respect to the simple world of an individual processes, where the notion of *halting forever* is not present.
- There is *downward causation*. The system of concurrently executing processes *constrains* the execution of an individual process by *indirect communication channels* established by the semaphore variables.

It is important to note that although this phenomenon is *fully explainable* it is *not predictable*, even in theory. If two processes try to execute the same semaphore operation exactly simultaneously, the underlying hardware enters into a state of *meta-stability* [33] (p. 77). It is not predictable, *even in theory,* which one of the two *simultaneous* processes will win this race.

It is also revealing to look at the problem of deadlock from the point of view of determinism. Although each one of the individual processes, the parts, behaves *deterministically* the behavior of the overall system, the whole, is *non-deterministic*.

5.2 Distributed Fault-Tolerant Clock Synchronization

In a time-triggered distributed computer system computational and communication processes are triggered by the progression of a global notion of physical time. This global notion of physical time must be *fault-tolerant* in order to mitigate the effects of a failing physical clock.

A distributed fault-tolerant synchronization algorithm constructs the fault-tolerant global time. Such an algorithm comprises the following three phases [22] (p. 69):

1. Periodic exchange of the time value of the local clock of each computing node among all the nodes of the system.
2. Distributed calculation of a global fault-tolerant time value, taking the local readings of the clock as inputs.
3. Adjustment of the local clock to come into agreement with the calculated global fault tolerant time value.

According to the theory of clock synchronization the number N of clocks in a system must be larger than $3\ k$, where k is the number of faulty clocks i.e., $N \geq (3k + 1)$.

A physical clock is a device that contains a physical oscillator (e.g., a crystal) and a counter that counts the number of ticks of the oscillator and thus contains the *state of the clock*. The frequency of the physical oscillator is determined by the *laws of physics* and depends on the size of the crystal and environmental conditions, such as temperature or pressure—a case of *upward causation*. The speed of the oscillator cannot be modified by *downward causation*. However, the state of the clock is modified by *downward causation* in step iii of the algorithm.

The phenomenon fault-tolerant clock synchronization fulfills the requirement of an emergent phenomenon:

- The phenomenon *fault-tolerant time, which does not fail if a single clock fails,* is novel with respect to the behavior of a single clock that can fail.
- There is *downward causation*. The system of concurrently executing clocks constrains the execution of an individual clock by adjusting the state of the counter of the local clock to a value that has been determined by the ensemble of clocks.

This example of emergence is interesting from the point of view of how *upward causation* (the frequency of a physical clock) and *downward causation* (the periodic correction of the state of a clock caused by the time value calculated by the ensemble of clocks at the macro level) interact and form a causal loop.

5.3 Alarm Processing

In an industrial plant an *alarm* is triggered when the value of a significant state variable exceeds a preset threshold limit. There may be thousands of significant state variables that are monitored in a large industrial plant. Since a single serious fault may cause a *correlated alarm shower* an alarm processing system must reduce the alarm rate at the operator interface to a manageable level in order to avoid an operator overload. The alarm processing system establishes the causal dependencies of alarms and decides which alarms can be hidden from the operator.

An alarm processing system consists of distributed sensors that can detect alarms and send alarm messages, a communication system that transports the alarm messages to an alarm processing center and the alarm analysis software that decides which alarm to hide.

Alarms are events that happen infrequently in normal operation. Many communication protocols for the transport of the alarm messages are of the PAR (Positive Acknowledgment of Retransmission) type for the transmission of event messages. The PAR protocol contains a retransmission mechanism to resend a message in case the previously sent message is not acknowledged in due time. Under heavy load, this mechanism can lead to a cascade effect

In the case of a correlated alarm shower that arises from a single serious fault, the event-triggered communication system slows down because the increased load on a finite capacity channel causes a delay of some messages. This slow-down induces the retransmission mechanism to kick in and to increase the load on the communication system even further. This can lead to a collapse called *thrashing*—an emergent phenomenon.

- The phenomenon *thrashing,* is novel with respect to the behavior under normal operation.
- There is *downward causation.* The high-load on the communication causes a slowdown of the communication system that causes the retransmission mechanism to increase the load even further.

5.4 Conway's Game of Life

Conway's Game of Life is a simple cellular automaton. It is played on a set of cells organized in a square array. Since there are no *things* involved, there is no upward causation from natural laws.

The simple rules of Conway's game of life are shown in Fig. 5. A player can select the initial conditions, i.e. the initial marking of the cells on the square array, as he/she pleases. After a round of updating all cells according to the transition rules, a new marking on the square array comes into sight. This marking forms the initial conditions for the following round, etc. Given defined initial condition, the series of states that develop is deterministic.

Let us choose the pattern for the initial conditions as shown in the left upper corner of Fig. 5. If all other cells of the square array are empty, then a phenomenon called *glider* appears.

If we select a *grain of observation* that observes the evolving patterns on the square array only after every four rounds then we clearly see the *glider* moving down diagonally along the square array. Holland calls this an emergent phenomenon [16].

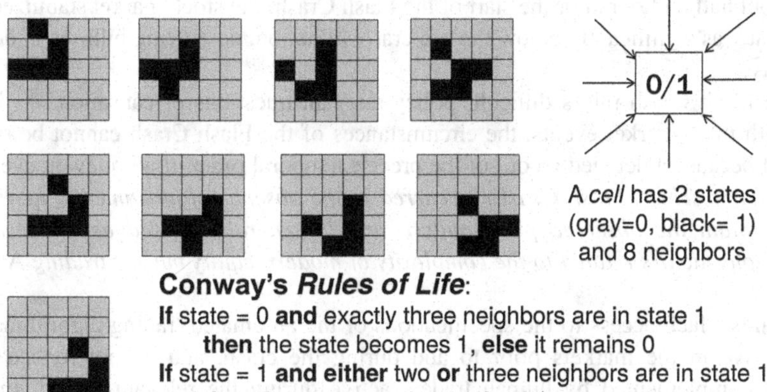

0/1
A *cell* has 2 states
(gray=0, black= 1)
and 8 neighbors
Conway's *Rules of Life*:
If state = 0 **and** exactly three neighbors are in state 1
then the state becomes 1, **else** it remains 0
If state = 1 **and either** two **or** three neighbors are in state 1
then the state remains 1, **else** it becomes 0
After four cycles, the pattern has moved along the diagonal.

Fig. 5. Conway's game of life

- The *moving glider* is a *deterministic consequence* of the selected initial conditions and the rules of the *game of life* at the micro-level. If the moving glider meets on its passage a *non-empty cell* of the square array then the *moving glider* disappears.
- The phenomenon of the *moving glider* that is observable on the selected macro level of a description hierarchy (Sect. 3.2) is *novel* and a surprise to a human observer. It is very difficult for the human mind to predict the patterns that will evolve deterministically form an initial condition in the course of many rounds.
- There is *downward causation (a feedback loop)* from one round to the next round, because the pattern that comes to sight after *all cells* have executed a round forms the initial condition *for each cell* in the following round.

5.5 Stock Market Crash on May 6, 2010

In today's electronic financial markets, an electronic trader can execute more than 1000 trades in a single second. The actions of a multitude of human traders and automated trading systems at the micro-level cause the valuation of the assets at the macro level which in turn influences the actions of the human traders and the algorithms of the automated trading systems, thus forming causal loops and cascade effects that can result in emergent misbehavior.

Aldrich et al. [2] reports about such a misbehavior of the stock market, called the Flash Crash on May 6, 2010: "… in the span of a mere four and half minutes, the Dow Jones Industrial Average lost approximately 1,000 points."

"As computerized high-frequency traders exited the stock market, the resulting lack of liquidity causes shares of some prominent companies to trade down as low as a penny or as high as $100.000" (N.Y Times, October 1, 2010)

About half an hour after the start of the Flash Crash, the stock market stabilized at a level that was significantly below the pre-crash valuation, destroying billions of dollars of equity.

The Flash Crash raises difficult, policy-relevant questions of causation. As is the case with most market events, the circumstances of the Flash Crash cannot be reconstructed because a detailed record of the precise temporal order of all relevant events is not available. *This "Flash Crash" occurred in the absence of fundamental news that could explain the observed price pattern and is generally viewed as the result of endogenous factors related to the complexity of modern equity market trading* Aldrich et al. [2].

Analysts lack access to the specifications of the automated trading algorithms that were active in the markets prior to and during the crash, and cannot replicate the strategies implemented by human traders active during the relevant period. Intense investigations and congressional hearings followed, but conclusive evidence is still missing six years after the crash. Although the sequence of events that caused the Flash Crash is explainable in theory it cannot be reconstructed in practice due to the concurrency and ignorance about the immense multitude of interacting transactions.

6 Consequences for CPSos Design

In CPSoS design not all the combinations allowed by Fig. 3 are of interest, in fact we are particularly interested in the behavior domain, i.e., behavioral emergence. Figure 6 classifies the emergent behavior of a CPSoS from the point of view of the consequences of this behavior on the overall mission of a CPSoS and from the prediction or awareness we may have on the appearance of emergent behavior.

Expected and beneficial emergent behavior is the normal case (quadrant 1) that results from a conscious design effort. *Unexpected and beneficial emergent* behavior is a positive surprise (quadrant 3). *Expected detrimental emergent* behavior can be avoided by adhering to proper design rules (quadrant 2). The problematic case is quadrant 4, *unexpected detrimental emergent behavior.*

In safety-critical CPSoSs, an unexpected detrimental emergent behavior can be the cause of a catastrophic accident. But how can we detect and avoid an *unknown* and therefore *unexpected* emergent phenomenon?

	Beneficial	Detrimental
Expected	1 Normal Case	2 Avoided by Design Rules
Unexpected	3 Positive Surprise	4 Problematic Case

Fig. 6. Contribution of emergent behavior

Clearly a conscious and aware design discipline aims to move, as knowledge progresses, more and more emergent phenomena from quadrant 4 to quadrant 2, in which provisions can be taken to mitigate, eliminate or prevent detrimental emergence. To exemplify just observe that while at its first manifestation deadlock was a problematic issue in distributed systems, today every computer student is though many of the different ways we have developed to properly address it.

Still our knowledge regarding CPSoS may remain limited and our ignorance about them can hardly be sufficiently reduced especially when we consider COTS components and legacy constituent systems. In fact, most CPSoS are built incorporating such LEGACY and COTS on which very little is known and where the information flow is often quite hidden.

In the remainder of this section we will focus on quadrant 4, the problematic case of *detrimental unexpected emergent with special* regards to undiscovered emergent phenomena never seen before.

6.1 Exposure of the Direct and Indirect Information Flow

In a CPSoS emergent behavior is the result of direct or indirect *flow of information* among the constituent systems.

At design time, the planned message-based, stigmergic and sometimes human information flow patterns should be analyzed in order to find *potential causal loops* and *cascade effects*. However, this analysis has limits where part of the information flow is hidden behind the interface of a CS whose interface model is incomplete because it abstracts from the details of the world behind the interface.

At run time, the actual information flow should be observed without the probe effect and documented with precise timestamps such that the temporal order of events can be reconstructed in a *post hoc* analysis of a scenario to establish the precise sequence that led to detrimental emergent behavior. This POST MORTEM analysis would be particularly useful to discover and explain new (just encountered) emergent phenomena. Actually such analysis, coupled with disclosure of the internal algorithms used for automatic trading would have allowed to explain the Stock Market Crash (Sect. 5.5).

6.2 Safety-Critical Systems

The behavior of a safety-critical system should conform to the *design model* that is the basis of the safety argument. The design model *does not* and *cannot* take into account *unknown emergent effects* that can cause a deviation of the actual behavior from the intended behavior.

Since in safety-critical CPSoS even a very small probability for a detrimental emergent phenomenon cannot be tolerated, it is proposed that the evolving state of a safety-critical CPSoS is meticulously monitored by an independent monitor component in order to detect the *onset* of an unexpected deviation of *the actual state* from the *intended state*. This deviation can be an indication for the start of an unknown (and therefore unexpected) detrimental emergent behavior. The system internal information flow to the monitoring system must operate in real-time in order that the monitor can act promptly. Since emergent behavior is *diachronic*, (i.e. it develops over time) an independent *meta (monitoring) system* that continually observes the evolving state of the *object system* can detect the early onset of a deviation and thus provide an immediate warning of a forthcoming disruption due to an emergent phenomenon. Based on this immediate warning, mitigating actions can be activated that bring the object system *back to normal operation* or at least to a *safe state.*

It is important to note that the monitoring system should be *state-based*, and not *process-based.* A state-based monitoring system acts on a higher-level of abstraction than a process-based system since it is concerned with the *properties of the states* of a system only and not with the much more involved processes that *generate the state changes.* A state-based monitoring system is thus much simpler than a process-based monitoring system. This fundamental difference between a state based and a process-based system is also important from the point of view of *design diversity* to detect hidden software errors.

Taking again the example of the Stock Market Crash (Sect. 5.5), if an independent monitoring system (without knowledge of the trading algorithms) had continually observed significant parameters that are relevant indicators of the market state and it had acted in the sub-millisecond range to stop the trading activities (safe state) the flash-crash that disrupted the market and wiped out billions of dollars of equity could have been avoided.

7 Conclusions

The purpose of building a Cyber-Physical System-of-Systems out of Constituent Systems (CSs) is to realize new services that go beyond the services provided by any of the CSs in isolation. *Emergence is thus at the core of CPSoS engineering.* In this Chapter we have surveyed some of the abundant past literature on emergence from the fields of philosophy and computer science, looked at the characteristics of multi-level hierarchies, developed a CPSoS definition of emergence and analyzed some examples of emergent behavior in computer systems.

We identified the basic mechanism that can lead to emergent phenomena: *causal loops* between the macro-level and the micro-level of a multi-level hierarchy (with the variant of *cascade effects*) that result in conceptually novel phenomena. We came to the conclusion that due to the ignorance about the scope of CPSoS even a thorough design analysis cannot uncover all potential mechanisms that can result in unexpected emergent phenomena at run-time. Unexpected emergent phenomena manifest themselves in a CPSoS by a diachronic deviation of the actual behavior from the intended (design) behavior.

Since *unknown emergent effects* can be the cause of a deviation of the actual behavior from the intended behavior, the meticulous observation of the behavior of a safety-critical CPSoS by an independent monitoring system can detect the onset of diachronic emergence and initiate mitigating actions before the detrimental emergent phenomenon has fully developed.

References

1. Ahl, V., Allen, T.F.H.: Hierarchy Theory. A Vision. Vocabulary and Epistemology. Columbia University Press, New York (1996)
2. Aldrich, E.M., Santa Cruz, U.C., Grundfest, J.: Stanford university law school, Laughlin, G., Santa Cruz, U.C.: The flash crash: a new deconstruction, 25 January 2016, Revised 2 February 2016. http://papers.ssrn.com/sol3/papers.cfm?abstract_id=2721922
3. Beckerman, A., et al. (eds.): Emergence or Reduction—Essays on the Progress of Nonreductive Physicalism. Walter de Gruyter, Berlin (1992)
4. Bedau, M.A., Humphreys, P.: Emergence, Contemporary Readings in Philosophy and Science. MIT Press, Cambridge (2008)
5. Boschetti, F., Gray, R.: Emergence and computability. ECO **9**(1), 120–130 (2007)
6. Clayton, P., Davies, P.: The Reemergence of Emergence. Oxford University Press, New York (2006)
7. Coffman, E.G., et al.: System deadlocks. ACM Comput. Surv. **3**(2), 67–78 (1971)
8. Fisher, D.A.: An emergent perspective on interoperation of system-of-systems. Technical report CMU/SEI-2006-TR-003, Carnegie Mellon University (2006)
9. Fromm, J.: The Emergence of Complexity. Kassel University Press, Kassel (2004)
10. Fromm, J.: Types and forms of emergence. arXiv.org/pdf//nlin/0506028.pdf. Accessed 12 May 2016
11. Gandolfi, K., Mourtel, C., Olivier, F.: Electromagnetic analysis: concrete results. In: Koç, Ç. K., Naccache, D., Paar, C. (eds.) CHES 2001. LNCS, vol. 2162, pp. 251–261. Springer, Heidelberg (2001). doi:10.1007/3-540-44709-1_21
12. Gligor, V.: Security of emergent properties in ad-hoc networks (transcript of discussion). In: Christianson, B., Crispo, B., Malcolm, J.A., Roe, M. (eds.) Security Protocols 2004. LNCS, vol. 3957, pp. 256–266. Springer, Heidelberg (2006). doi:10.1007/11861386_30
13. Hempel, C.B., Oppenheim, P.: Studies in the logic of explanation. Philos. Sci. **15**(2), 135–175 (1948)
14. Henshaw, M., et al.: The Systems of systems engineering strategic research angenda. Document Number: TAREA-PU-WP5-R-LU-26. Issue 1, Loughborough University, United Kingdom 3, 17 June 2013
15. Hoeftberger, O.: Report on the AMADEOS Workshop on Emergence in Cyber Physical-Systems of Systems. Vienna University of Technology, May 2016
16. Holland, J.H.: Emergence, from Chaos to Order. Oxford University Press, New York (1998)
17. Huberman, B.A., Hogg, T.: The Ecology of Computation. Studies in Computer Science and Artificial Intelligence, vol. 2, pp. 73–115. North Holland, Amsterdam (1988)
18. Jamshidi, M.: Systems of Systems Engineering. Wiley, Hoboken (2009)
19. Keating, C.H.: Research foundations for systems of systems engineering. In: Proceedings of the International Conference on Systems, Man and Cybernetics, vol. 3, pp. 2720–2725 (2005)
20. Kim, J.: Emergence: core ideas and issues, 9 August 2006. http://cs.calstatela.edu/~wiki/images/b/b1/Emergence-_Coreideas_and_issues.pdf
21. Koestler, A.: The Ghost in the Machine. Hutchinson, London (1967)
22. Kopetz, H.: Real-time Systems-Design Principles for Distributed Embedded Applications. Springer, Heidelberg (2011)
23. Kopetz, H.: A conceptual model for the information transfer in systems of systems. In: Proceedings of the 17th ISORC, pp. 17–24. IEEE Press (2014)
24. Kopetz, H., et al.: Direct versus stigmergic information flow in systems-of-systems. In: Proceedings of SoSE 2015, pp. 36–41. IEEE Press (2015)

25. McLaughlin, B., Bennet, K.: Supervenience. Stanford Encyclopedia of Philosphy, Stanford (2011)
26. Mogul, J.: Emergent (Mis)behavior vs. Complex Software Systems. In: Proceedings of EuroSys, pp. 293–304 (2006)
27. O'Connor, T.: Emergent Properties. Stanford Encyclopedia of Philosophy, Stanford (2012)
28. Parunak, H.V., VanderBok, R.S.: Managing emergent behavior in distributed control systems. In: Proceedings of ISA Tech 1997, Anaheim, CA (1997)
29. Pattee, H.H.: Causation, control, and the evolution of complexity. In: Anderson, P.B., et al. (eds.) From: Downward Causation: Mind, Bodies, Matter, pp. 63–77. Aarhus University Press, Aarhus (2000)
30. Pattee, H.H.: The physical basis and origin of hierarchical control. In: Pattee, H.H. (ed.) Laws, Language and Life, vol. 7, pp. 91–110. Springer, New York (2012)
31. Ronal, E.M.A., et al.: Design, observation, surprise! a test of emergence. Artif. Life **5**, 225–239 (1999)
32. Simon, H.: The architecture of complexity. In: Simon, H. (ed.) The Science of the Artificial. MIT Press, Cambridge (1969)
33. Sparso, J., Furber, S.: Principles of Asynchronous Circuit Design. Kluwer Publisher, Dordrecht (2002)
34. Stephan, A.: Emergence–a systematic view on its historical facets. In: Beckerman, E., et al. (eds.) Emergence or Reduction?, pp. 25–48. Walter de Gruter, Berlin (1992)

AMADEOS SysML Profile for SoS Conceptual Modeling

Paolo Lollini[1(✉)], Marco Mori[1], Arun Babu[2], and Sara Bouchenak[3]

[1] Department of Mathematics and Informatics,
University of Florence, Firenze, Italy
{paolo.lollini,marco.mori}@unifi.it
[2] Resiltech SRL, Pisa, Italy
arun.babu@resiltech.com
[3] Université Grenoble Alpes, Grenoble, France
sara.bouchenak@insa-lyon.fr

1 Introduction

In the European Union FP7-610535-AMADEOS project, a conceptual model for Systems of Systems (SoSs) has been conceived to find a common language allowing experts to collaborate on modelling, engineering, and analyzing SoSs (see public deliverable D2.3 "AMADEOS conceptual model - Revised" [1]).

Analogously to the conceptual model for the architecture of software intensive systems, we separated the description of basics SoS concepts into different perspectives. These perspectives are called viewpoints, each of which is focused on different concerns of the SoS: *structure*, *evolution*, *dynamicity*, *dependability, security*, *time*, *multi-criticality* and *emergence*.

- *Structure*: It represents architectural concerns of an SoS. In particular it defines the manner in which Constituent Systems (CSs) are composed [17] and how do they exchange semantically well-defined messages [10] through their interfaces [22].
- *Evolution and dynamicity*: Dynamicity represents variations to the operation of SoS that have been considered at design-time to reconfigure the SoS in specific situations e.g., either after a fault or after the variation of an external condition [21]. Evolution represents changes that have been introduced later to accommodate modified or new requirements by means of including, removing or modifying system functions [16].
- *Dependability* and *security* [2]: It consists of non-functional critical requirements as availability, reliability, safety, privacy or confidentiality.
- *Time*: It is fundamental since SoSs are sensitive to the progression of time and it is necessary to design responsive SoSs able to achieve reliably time-dependent requirements [9].

This work has been partially supported by the FP7-610535-AMADEOS project.

A. Bondavalli et al. (Eds.): Cyber-Physical Systems of Systems, LNCS 10099, pp. 97–127, 2016.
DOI: 10.1007/978-3-319-47590-5_4

- *Multi-criticality*: It aims at integrating together subsystems providing services with different levels of criticality corresponding to different dependability and security requirements [23].
- *Emergence*: It mainly denotes the appearance of novel phenomena at the SoS level that are not observable at CSs level; managing *emergence* is essential to avoid undesired, possibly unexpected, situations generated from CSs interactions as well as to realize desired emergent phenomena being usually the higher goal of an SoS [14].

In this chapter we will focus on the basic SoS concepts belonging to the different viewpoints and on their semantic relationships, and we will present a SysML profile to represent the conceptual model.

The rest of this chapter is structured as follows: Sect. 2 presents the different concepts defined in a SysML profile to model an SoS. Section 3 describes the structural properties of an SoS in term of architecture, communication and interface. Section 4 defines the concept of evolution related to all changes of an SoS. Section 5 presents the concept of dynamicity that represents the variation to the operation of an SoS considered at design time. Section 6 describes the concepts related dependability, security and multi-criticality aspects. Section 7 describes the global notion of time exploited in an SoS, while Sect. 8 defines the concept of emergence of novel phenomena at the SoS level. Then, Sect. 9 introduces a concrete case study to illustrate the application of basic SoS concepts. Lastly, Sect. 10 provides a brief overview of related works before the conclusion in Sect. 11.

2 Conceptual Modeling Support: The AMADEOS SysML Profile

This section focuses on the definition of a SysML profile as a modeling support for representing the basic concepts for SoS and their relationships. Following the viewpoint-driven approach previously introduced, the concepts and their relationships have been modeled using a SysML semi-formal representation, organized in a profile[1] composed by viewpoint-related packages. To this end, we have defined specific constructs and we have exploited already implemented stereotypes available in other related profiles to support specific viewpoints. Our proposed profile is meant to be used by designers in describing the static SoS structure and its dynamic behavior according to the introduced viewpoints. Such an SoS description can be adopted to be kept consistent across viewpoints by tools and for machine-assisted cross-viewpoint analyses (e.g., finding detrimental emergent SoS behavior).

The SoS profile will be used as an abstract model to represent the topology and the state evolution of an operational SoS. The profile diagrams contain the SoS basic concepts distributed in sub-packages as follows:

[1] https://github.com/AMADEOSConceptualModel/SysMLProfileAndApplication.git - GitHub public link to the AMADEOS SysML profile and the Smart Grid application.

- *SoS Architecture*: describes the basic architectural elements and their semantic relationships.
- *SoS Communication*: provides the fundamental elements in order to describe the behavior of an SoS in terms of sequence of messages exchanged among CSs.
- *SoS Interface*: describes all the points of integration that allow the exchange of information among the connected entities.
- *SoS Dependability*: provides the basic concepts related to SoS dependability.
- *SoS Security*: provides the basic concepts related to SoS security.
- *SoS Evolution*: provides the main elements to describe the process of gradual and progressive change of an SoS.
- *SoS Dynamicity*: provides basic concepts related to SoS dynamicity.
- *SoS Scenario-based reasoning*: provides the basic concepts for supporting the generation, evaluation and management of different scenarios resulting from SoS dynamicity, thus supporting decision-making in an SoS.
- *SoS Time*: provides the fundamental elements to describe time concepts.
- *SoS Multi-Criticality*: provide the basic concepts to describe the multi-criticality aspects of an SoS.
- *SoS Emergence*: provides the main elements to describe the SoS emergence concepts.

It is worth noticing that most of the above packages come from a direct mapping to the views previously defined except for *SoS Architecture, SoS Communication* and *SoS Interface* that all together implement the Structure view, and for *SoS Dynamicity* and *SoS Scenario-based reasoning* that map into the Dynamicity view.

We have implemented the whole profile by exploiting the Eclipse integrated development environment, jointly with Papyrus. Eclipse is an open source environment and offers all the related advantages in terms of cost, customizability, flexibility and interoperability. Papyrus is an Eclipse plugin, which offers a very advanced support to define UML profiles.

In the following sections, we will discuss the key elements of the conceptual model for each identified viewpoint. All the new introduced stereotypes extend the "Block" stereotype of SysML, if not differently specified. For the sake of readability, we will not represent such relations in the SysML diagrams describing the different packages.

3 Structure Viewpoint

The viewpoint of *structure* represents architectural concerns of an SoS. In particular, it defines the manner in which CSs are composed [17] and how do they exchange semantically well-defined messages [10] through their interfaces [22].

The static structure of an SoS is based on the concept of a *Constituent System (CS)*, which is *'An autonomous subsystem of an SoS, consisting of computer systems and possibly of a controlled objects and/or human role players that interact to provide a given service'*. A CS exchanges *information* that is either represented by things/energy or data with its *environment* by means of *interfaces*. The environment of a CS includes all entities that are able to interact with the CS, including other CSs. In our context, information is a proposition about the state of or an action in the world, which is either an attribute of a physical thing (e.g., temperature of a room) or an attribute of an abstract construct (e.g., execution time of a program).

The interfaces among which the CSs interact one another are the *Relied Upon Interfaces (RUIs)*. As such, the CS *service* – which is its intended behavior – is provided at this interface. RUI is further structured in the *Relied Upon Message Interface (RUMI)* and the *Relied Upon Physical Interface (RUPI)*. RUMI allows for message-based communication of CSs over cyberspace (e.g., the Internet) while RUPI enables the indirect physical exchange of things or energy among CSs over their common environment. It consists of sensors and actuators that take and time-stamp observations of and/or act at a defined deadline on some physical state (e.g., the temperature of a room) in the physical environment according to their design. Environmental dynamics (e.g., heat dissipation through walls) act additionally to other CSs on the physical state. CSs that interact with each other over a common physical environment establish a *stigmergic* channel, i.e., they communicate indirectly over influencing and measuring the physical state. For more details on the interface topic, please refer to [11], and Chapter 2 of this book.

The profile supports the description of the static and dynamic structure of an SoS representing: the basic architectural elements and their semantic relationships; the sequence of messages exchanged among CSs in an SoS; the points of integration, i.e., interfaces, allowing the exchange of information/energy among connected entities.

The structural properties of an SoS are described using three different packages "*SoS Architecture*" (Sect. 3.1), "*SoS Communication*" (Sect. 3.2), and "*SoS Interface*" (Sect. 3.3). The first defines Stereotypes useful to describe the topology of an SoS; the second provides Stereotypes to describe the communication aspects between the Constituent Systems of an SoS; finally, "*SoS Interface*" semi-formalizes internal and external points of interaction of an SoS.

3.1 SoS Architecture Package

Architectural components are defined within the "*SoS Architecture*" package (see Fig. 1). This package extends SysML Block Definition Diagram (BDD) in order to model the topology and the relations of an SoS. Blocks in SysML BDD are the basic structural element used to model the structure of systems (Wolfrom) and they can be used to represent: systems, system components (hardware and software), items, conceptual entities and logical abstractions. A Block is depicted as a rectangle with compartments that contain Block characteristics such as: name, properties, operations and requirements that the Block satisfies. A Block provides a unifying concept to describe the structure of an element or a system: System, Hardware, Software, Data, Procedure, Facility and Person.

This type of diagram helps a system designer to depict the static structure of an SoS in terms of its constituent system and possible relationships.

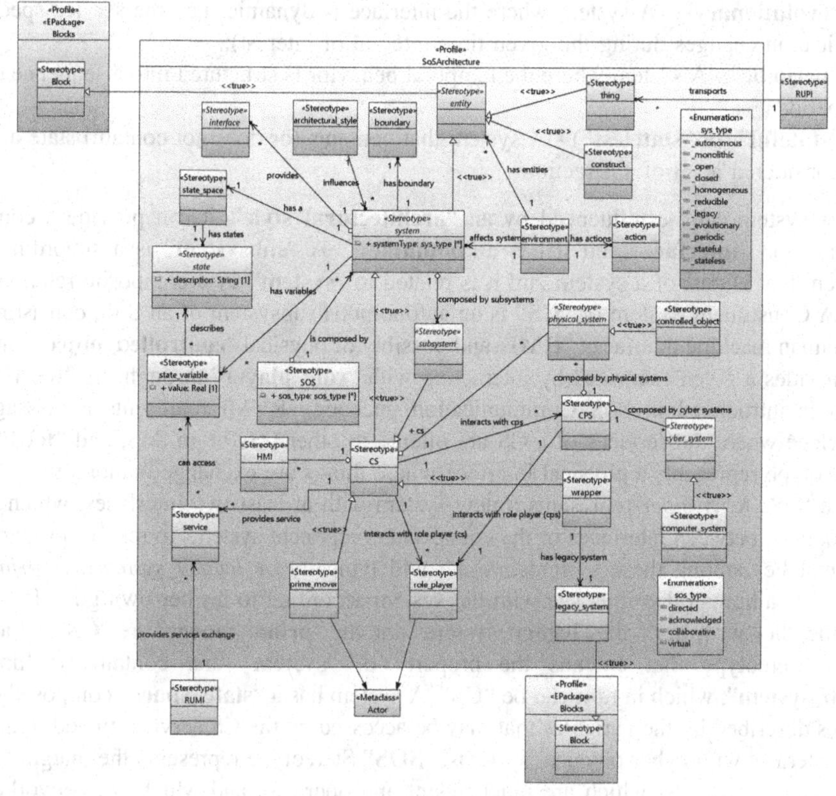

Fig. 1. SoS Architecture package

The first Stereotype is "**entity**" and it extends the SysML metaclass "Block". We distinguish between two different kinds of entities: "**thing**" or "**construct**". They extend the properties of "entity" and so they are also represented as Blocks.

A "**System**" is a type of entity (thereby a Block), it has the same characteristic but it is also capable of interacting with its environment. As it is expressed by the "**sys_type**" Enumeration, a system can be:

- "**autonomous**" - A system that can provide its services without guidance by another system;
- "**monolithic**" - if distinguishable services are not clearly separated in the implementation but are interwoven;
- "**open**" (or "**closed**") - A system that is interacting (or is not interacting) with its environment during the given time interval of interest;
- "**legacy**" - An existing operational system within an organization that provides an indispensable service to the organization;
- "**homogeneous**" - A system where all sub-systems adhere to the same architectural style;
- "**reducible**" - A system where the sum of the parts makes the whole;

- "**evolutionary**" - A system where the interface is dynamic (i.e., the service specification changes during the given time interval of interest);
- "**periodic**" - A system where the temporal behavior is structured into a sequence of periods.
- "**stateful**" (or "**stateless**") - A system that contains (or does not contain) state at a considered level of abstraction.

A system can be influenced by an "**architectural_style**", it can provide a communication "**interface**" and it has a "**boundary**". A "**subsystem**" is a subordinate system that is part of a system and it is related to "**system**" by a composite relation.

A Constituent System or "**CS**" is an autonomous subsystem of an SoS, consisting of human machine interfaces "**HMI**" and possibly of physical "**controlled_object**" and it provides a given "**service**" by interacting with "**role_player**" through the "**RUMI**" (that is introduced in SoS Communication package). RUMI represents a message interface where the services of a CS are offered to other CSs of an SoS, and "**RUPI**" Stereotype represents a physical interface where things are exchanged among the CSs of an SoS. A *wrapper* represents a new system with at least two interfaces, which is introduced between interfaces of the connected component systems to resolve property mismatches among these systems, which will typically be *legacy_systems*. A *prime mover* is a human that interacts with the system according to his/her own goal. In the profile, the "**wrapper**", the "**legacy_system**" and the "**prime_mover**" are "**CS**", which is a Stereotype that extends the property of "**system**" that contains multiple "**sub_system**", which in turn can be "**CS**". A system has a "**state_space**" composed of states described by the variables that may be accessed by the CS service. In addition, a CS interacts with cyber-physical systems. "**SOS**" Stereotype represents the integration of systems, i.e., CSs which are independent and operable, and which are networked together for a period of time to achieve a certain goal. As expressed by the "**sos_type**" Enumeration, an SoS can be:

- "**directed**" - An SoS with a central managed purpose and central ownership of all CSs;
- "**acknowledged**" - Independent ownership of the CSs, but cooperative agreements among the owners to an aligned purpose;
- "**collaborative**" - Voluntary interactions of independent CSs to achieve a goal that is beneficial to the individual CS;
- "**virtual**" - Lack of central purpose and central alignment.

A Cyber-Physical System ("**CPS**") is composed by a set of "**cyber_system**" (i.e., computer systems), and "**physical_system**" (i.e., controlled objects).

3.2 SoS Communication Package

The "*SoS Communication*" package (see Fig. 2) is composed of CSs that exchange information with other elements. In order to represent the exchanged information during the progression of time we use a SysML Sequence Diagram and we represent a CS not only as a Block entity of BDD but also as "Lifeline" metaclass. "Lifeline" is a metaclass

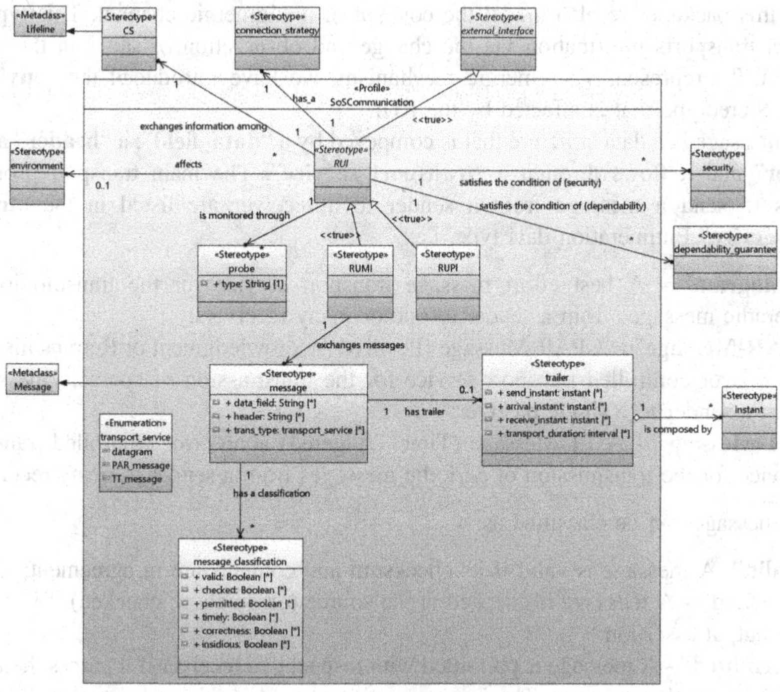

Fig. 2. SoS Communication package

and part of Sequence Diagrams. Through a Sequence Diagram it is possible to represent the behavior of a system in terms of a sequence of messages exchanged between parts and a "Lifeline" defines the individual participants in the interaction (Constituent System). Moreover, through a "Lifeline" it is possible to describe the temporal behavior of an SoS. The time is showed by the length of the "Lifeline" and it passes from top to bottom: the interaction starts near the top of the diagram and ends at the bottom.

A "**RUI**" Stereotype represents an external interface of a CS where the services of a CS are offered to other CSs. It extends "**external_interface**" (defined in "*SoS Interface*" package) and guarantees the exchange of information among CSs ("**CS**" is defined in "*SoS Architecture*" package). A RUI can be represented also as a Sequence Diagram in which CSs are represented by the lifelines that exchange information. A RUI, can be either a "**RUMI**" or a "**RUPI**" and it is monitored through "**probes**". A *RUI connecting strategy* is part of the interface specification that searches for desired, w.r.t. connections available, and compatible RUIs of other CSs and connects them until they either become *undesirable*, *unavailable*, or *incompatible*. A RUI, having a "**connection_strategy**", is instantiated complying to possibly multiple "**dependability_guarantees**" and satisfying "**security**" constraints.

A "**RUMI**" represents a message interface for the exchange of information among two or more CSs and extends the "**RUI**" Stereotype. While messages are exchanged through the RUMI, physical elements are exchanged among the CSs of an SoS through the "**RUPI**"; physical elements are things or energy.

In this package we also model the concept of a stigmergic channel. This type of channel transports information via the change and observation of states in the environment. To represent a stigmergic mechanism, we have introduced the "**environment**" Stereotype that is affected by the RUI.

A message is a data structure that is composed by a "**data_field**", a "**header**" and a "**trailer**" and it flows through a "**transport_service**". The main transport protocol classes to send a message from a sender to a receiver are listed in the "**transport_service**" Enumeration data type, i.e.:

- "**datagram**" - A best effort message transport service for the transmission of sporadic messages from a sender to one or many receivers;
- "**PAR-Message**" - A PAR-Message (Positive Acknowledgment or Retransmission) is an error controlled transport service for the transmission of sporadic messages from a sender to a single receiver;
- "**TT-Message**" - A TT-Message (Time-Triggered) is an error controlled transport service for the transmission of periodic messages from a sender to many receivers.

A message can be classified as:

- "**valid**"- A message is valid if its checksum and contents are in agreement;
- "**checked**" - A message is checked at the source (or, in short, checked) if it passes the output assertion;
- "**permitted**" - A message is permitted with respect to a receiver if it passes the input assertion of that receiver. The input assertion should verify, at least, that the message is valid;
- "**timely**" - A message is timely if it is in agreement with the temporal specification;
- "**correctness**" - A message is correct if it is both timely and value correct. A message is value-correct if it is in agreement with the value specification;
- "**insidious**" - A message is insidious if it is permitted but incorrect.

3.3 SoS Interface Package

The interfaces are the key issue to the integration of systems (see also Chap. 2 of this book) and in this section we introduce an in-depth analysis of the SoS interface concepts, which are represented in Fig. 3.

An interface can be an "**internal_interface**" a "**physical_interface**", a "**message_based_interface**" and an "**external_interface**". The internal interface connects two or more subsystems of a CS (the Stereotype "subsystem", defined in *SoS Architecture* package, is connected with "internal_interface" in order to represent this relation). The physical interface consists of three different types of elements, namely "**sensor**", "**actuator**" and "**transducer**". The "**message_based_interface**" allows the transmission of message by means of "**message**" which are defined in terms of "**message_variable**". Finally, the external interface connects two or more CS (the Stereotype "CS" is connected with "**external_interface**"). A different type of "**external_interface**" is the "**utility_interface**", which is an interface of a CS that is used for the configuration, or the control, or the observation of the behavior of the CS. The

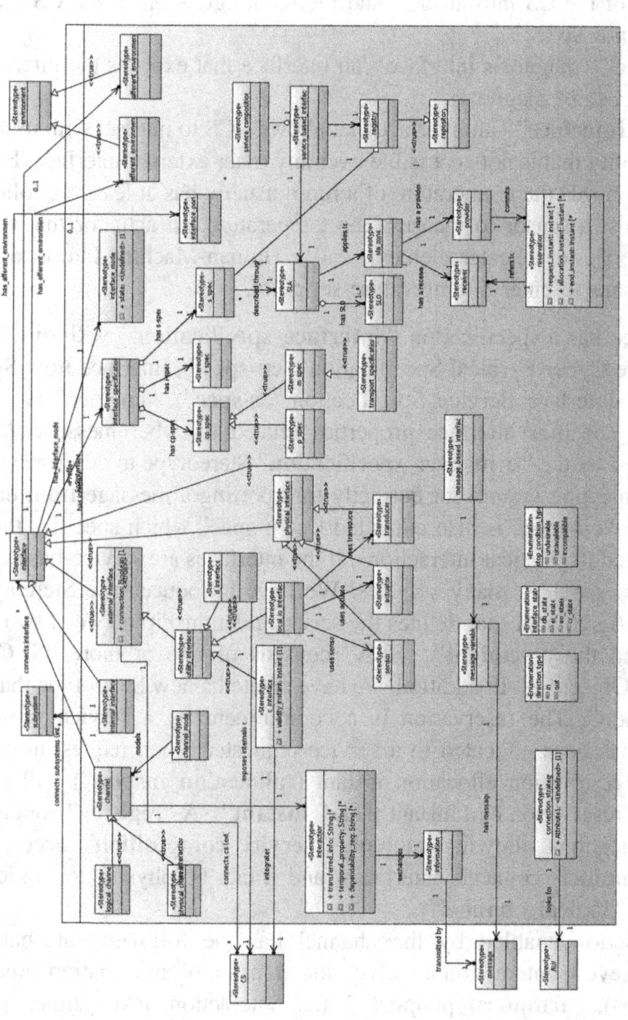

Fig. 3. SoS Interface package

purposes of the utility interfaces are to (i) configure and update a CS, (ii) diagnose a CS, and (iii) let a CS interact with its remaining local physical environment that is unrelated to the operative services of the SoS.

The utility interface is specialized into three different types of interfaces:

- "**c-interface**" - configuration interface - an interface of a CS that is used for the integration of the CS into an SoS and the reconfiguration of the CS's RUIs while integrated in a SoS.
- "**d-interface**" - diagnosis interface - an interface that exposes the internals of a CS for the purpose of diagnosis.
- "**local_IO_Interface**" - an interface that allows a CS to interact with its surrounding physical reality that is not accessible over any other external interface. For example, a CS that controls the temperature of a room usually has at least the following local IO Interfaces: a sensor to measure the temperature, an actuator that regulates the flow of energy to a heater element, and a Human-Machine-Interface (HMI) that allows humans to enter a temperature set point.

An interface has a specification ("**interface_specification**") with different kind of levels: Interface Cyber-Physical Specification ("**cp-spec**"), Interface Itom Specification ("**i-spec**") and Interface Service Specification ("**s-spec**"). "**cp-spec**" is extended by "**m-spec**" that specifies interface properties related to cyber message. "**m-spec**" is further extended by the "**transport_specification**" Stereotype to describe all properties of the communication system for correctly transporting a message from the sender to the receiver(s). "**cp-spec**" is also extended by "**p-spec**" which specifies the interfaces properties related to physical interactions. If the interfaces are service-based, this means that the system provides many services. We have introduced the Stereotype "**SLA**" Service Level Agreement that defines the service relationship between two parties: the "**provider**" and the "**recipient**". "**SLA**" consists of one or more "**SLO**", i.e., the Service Level Objectives. In addition, we have created a new Stereotype that represents the "**reservation**". The reservation is a commitment by a service provider that a resource that has been allocated to a service requester (upon request at "**request_instant**") at the reservation allocation instant ("**allocation_instant**") will remain allocated until the reservation end instant ("**end_instant**"). A "**registry**" contains multiple service specifications allowing multiple "**service_composition**" according to the "**SLA**". A "**channel**" connects interfaces, and it can be physical or logical ("**physical_channel**", "**logical_channel**").

The interaction enabled by the channel has the following attributes: "**transferred_info**" (every interaction involves the transfer of information among participating systems), "**temporal_property**" (an interaction takes time, i.e., for an interaction to occur it is initiated and completed according to system-specific temporal properties) and "**dependability_req**" (e.g., interactions might require resilience with respect to perturbation or need to guarantee security properties like confidentiality). Through channel interactions, the information is transmitted by means of messages". A "**channel_model**" describes the effects of the channel on the transferred information. An "**interface_model**" contains the explanation of the interface. An interface, associated to an "**interface_port**" has an afferent and an efferent "**interface_model**", which are affected and may affect the interface, respectively. A "**connection_strategy**" Stereotype is defined and connected to a RUI.

4 Evolution Viewpoint

Large scale Systems-of-Systems (SoSs) tend to be designed for a long period of usage (10 years+). Over time, the demands and the constraints put on the system will usually change, as will the environment in which the system is to operate. The AMADEOS project studied the design of systems of systems that are not just robust to dynamicity (short-term change), but to long-term changes as well. *Evolution* represents changes that have been introduced later to accommodate modified or new requirements by means of including, removing or modifying system functions [16].

In contrast to dynamicity, the concept of evolution relates to all changes of an SoS that are not given by design, but arise by changes in the environment (primary evolution), or by new or changed requirements on the SoS service itself (secondary evolution). In the prospect of formalizing a methodology that allows evolution to take place in a controlled manner, the concept of *managed evolution* is most relevant. It is defined as the *'evolution that is guided and supported to achieve a certain goal'* [16];

4.1 SoS Evolution Package

In order to describe this type of processes we have chosen a Block Definition Diagram, because it is designed to show the generic characteristics and structures of a system.

The main SoS concepts are modelled within the "*SoS Evolution*" package of our SoS profile. Figure 4 shows the "**evolution**" Stereotype as a Block of a BDD, aiming at describing an SoS change. In our conceptual model we envision two different types of evolution:

- "**managed_evolution**" - Process of modifying the SoS to keep it relevant in face of an ever-changing environment. Examples of environmental changes include new available technology, new business cases/strategies, new business processes, changing user needs, new legal requirements, compliance rules and safety regulations, changing political issues, new standards, etc.
- "**unmanaged_evolution**" - Ongoing modification of the SoS that occurs as a result of ongoing changes in (some of) its CSs. Examples of such internal changes include changing circumstances, ongoing optimization, etc.

An SoS evolution has a "**goal**", improves the "**business value**" by means of the exploit of "**system_resource**" and can be affected by the environment. Evolution is achieved by modifying CSs and consequently the whole SoS.

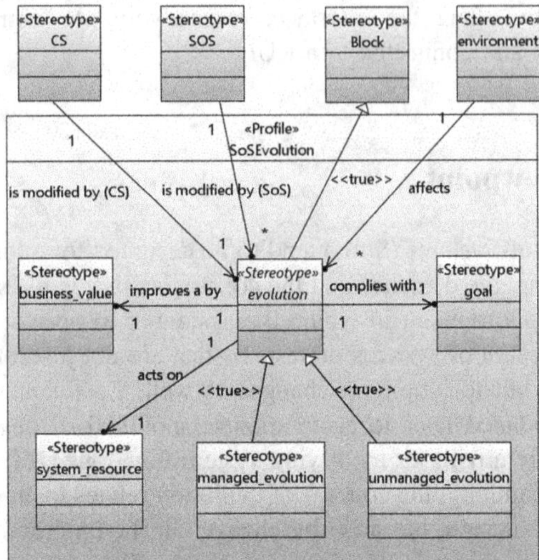

Fig. 4. SoS evolution package

5 Dynamicity Viewpoint

Dynamicity is the property of an entity that constantly changes in term of offered services, built-in structure and interactions with other entities. It represents variations to the operation of SoS that have been considered at design-time to reconfigure the SoS in specific situations e.g., either after a fault or after the variation of an external condition [21]. Dynamicity encompasses all interactions, e.g., message exchange over time.

Closely related to dynamicity is the concept of *reconfigurability*, which is the ability of a system to change its configuration according to the current demands.

The Dynamicity components are described by means of two different packages, i.e., "SoS Dynamicity" (Sect. 5.1) and "SoS Scenario-based reasoning" (Sect. 5.2).

5.1 SoS Dynamicity Package

In this section we show how to use a semi-formal language in order to represent the dynamicity of an SoS. Our objective is to (1) identify which parts of an SoS are dynamic at a certain extent and (2) to represent the dynamic behavior through the interactions among CSs.

As presented in Fig. 5 we have introduced the concept of **"dynamicity"** (belonging to the already defined stereotype **"entity"**), which can be applied either to a CS or to a whole SoS. Dynamicity may be of different nature, either **"dynamic_service"**, or **"reconfigurability"**, i.e., the variation to the CSs architecture, or **"dynamic_interaction"**. Already defined concepts like **"service"** and **"interaction"** are the objects of a dynamic behavior.

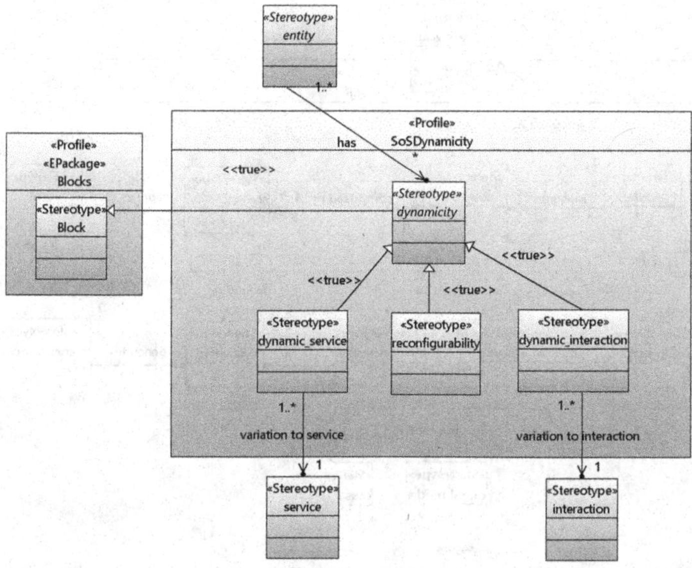

Fig. 5. SoS Dynamicity package

Eliciting dynamicity behavior of different nature that applies to different portions of an SoS is not enough to have a full understanding of the dynamic behavior. With this aim, along with the dynamicity package, we have considered interaction diagrams in order to focus on the message interchange between a number of lifelines: Sequence Diagrams. We propose a methodology to be used to represent dynamicity as it follows:

- Making use of Sequence Diagrams to represent the system behavior in terms of a sequence messages exchanged between parts;
- Selecting the constituent systems involved in the communication;
- Describing the most common interactions.

This type of representation helps a system designer to understand which are the properties of an SoS that are constantly changing and how the SoS can change and rearrange its components. The dynamic introduction, modification or removal of constituent systems can introduce new system behaviors that need to be analyzed.

5.2 SoS Scenario-Based Reasoning Package

Scenario based reasoning package aims at supporting dynamicity and evolution of an SoS. By means of this component of the profile we aim at supporting the generation, evaluation and management of different scenarios thus supporting decision-making in an SoS. As shown in Fig. 6, the main concept of this component is **"scenario"** which is composed by a set of **"scenario_state"** each of which associated to an **"event"** to be applied at each state. A state is in instantiation of a set of **"variables"** which are relevant for the decision-making. Such variables can be extracted by means of an

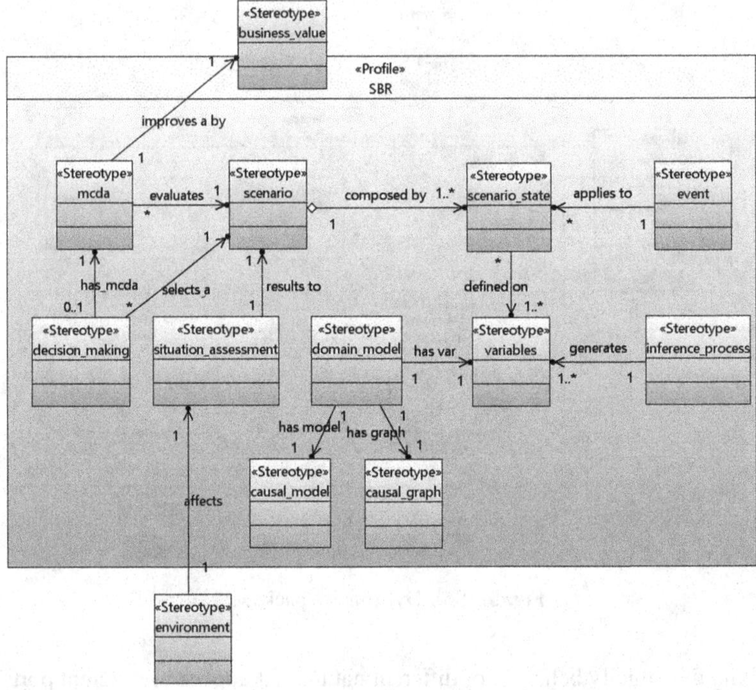

Fig. 6. SoS Scenario-Based Reasoning (SBR) package

"inference_process" and they pertain to a **"domain_model".** The latter defines relationships among variables in terms of correlations (**"causal_model"**) and causation (**"causal_graph"**) dependencies.

The process of generating scenario results from the **"situation assessment"** that depends on the **"environment"**. **"decision_making"** is the process to select a course of actions among different possible alternate scenarios. A multi-criteria decision analysis **"mcda"** may be also applied to improve the decision-making process. Finally, scenarios are subject to pruning and updating operations in order to discard non-correct or un-likely scenarios and to update scenarios dealing with newly available information.

6 Dependability, Security, and Multi-criticality Viewpoints

In any large system, faults and threats are normal and may impact on the availability, reliability, maintainability, safety, data integrity, data privacy, and confidentiality. Traditional dependability and security concepts [2] like fault, error and failure, have been included in the conceptual model. Dependability integrates the attributes of availability, reliability, maintainability, safety, integrity and robustness, and it can be attained by means of fault prevention, fault tolerance, fault removal and fault forecast.

Security is impacted by threats that impose risks exploiting possible SoS vulnerabilities. It is the composition of confidentiality, integrity, and availability; security requires in effect the concurrent existence of availability for authorized actions only, confidentiality, and integrity (with "improper" meaning "unauthorized").

Confidentiality is ensured by means of encryption. Keys are used for encryption/decryption operations, which can be public or private. In an access control system, the security policy is enforced by what is called the reference monitor, which represents the mechanism that implements the access control model. Authorization assigns permissions, which are defined in a security policy. A security policy relies on trusted systems, which encompass hardware, software or human components.

Multi-criticality aims at integrating together subsystems providing services with different levels of criticality corresponding to different dependability and security requirements [23].

A multi-critical SoS is a system containing several components that execute applications with different criticality, such as safety-critical and non-safety-critical. The architecture of safety-critical applications shall be built taking into account that while some part of the system may have strong safety-critical requirements, other parts may be not so critical.

For example, a railway system is a multi-criticality system, given that it consists of components that deliver services at different criticality levels, e.g., a braking service and a heating service. These components usually adhere to different Safety Integrity Levels (SIL) resulting in a system exhibiting different levels of criticality.

In the following we describe the three different packages supporting the definition of dependability (Sect. 6.1), security (Sect. 6.2) and multi-criticality (Sect. 6.3) aspects. The terminology is based on canonical definitions of dependability and security concerns as defined in [2].

6.1 SoS Dependability Package

Figure 7 shows the key concepts captured within the dependability package. A CS or a whole SoS may require possible multiple "**dependability_guarantee**" through the achievement of possible different dependability "**metric**" by means of possible different "**technique**".

A technique is exploited to reduce the occurrence of faults: "**fault_prevention**", "**fault_tolerance**", "**fault_removal**", "**fault_forecast**".

A "**measure**" represents a property expected from a dependable system expressed in terms of a quantitative "**target_value**": "**availability**", "**reliability**", "**maintainability**", "**safety**", "**integrity**", "**robustness**".

The profile supports the definition of "**fault_containment region**", "**error_ containment**" and "**error_containment_region**". The first contains components operating correctly regardless of any arbitrary fault outside the region. These components may have erroneous output actions that are alleviated with the definition of "**error_containment**", which prevents propagation of errors by employing error detection and a mitigation strategy. This leads to the definition of "**error_containment_region**" which contains more "**fault_containment region**" having "**error_containment**". A Fault

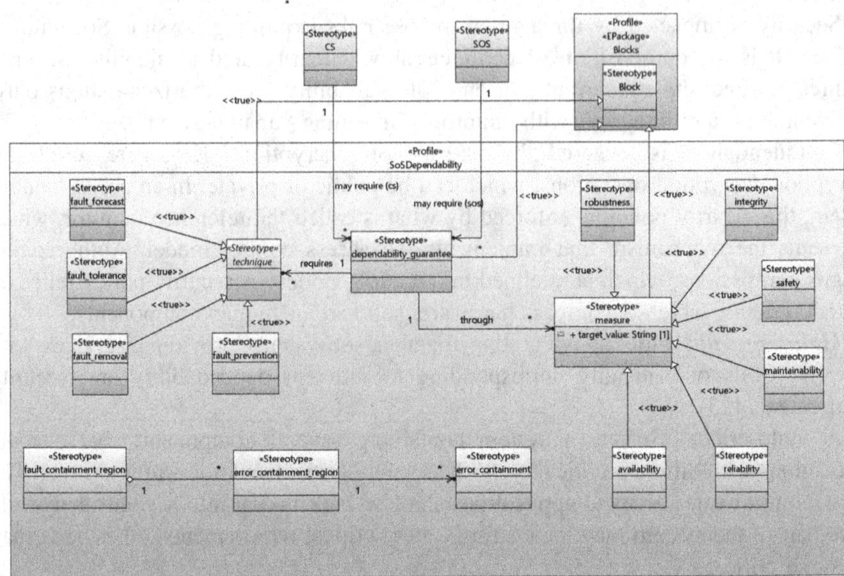

Fig. 7. Dependability conceptual model.

Containment Region (FCR) is a collection of components that operates correctly regardless of any arbitrary fault outside the region.

6.2 SoS Security Package

This section describes the fundamental elements used by a system designer to represent security aspects of an SoS.

As shown in Fig. 8, we connect the Stereotype "**SOS**" and "**CS**" to "**security**" Stereotype to satisfy the security conditions of an SoS. To this end we use "**cryptography**" based on symmetric ("**symmetric_cryptography**") or public key ("**public_key_cryptography**") infrastructure.

The "**encryption**" Stereotype represents the process of disguising data in such a way to hide the information it contains. In this way, data exchanged between Constituent Systems are processed using a cryptography key. Three types of key have been represented: "**symmetric_key**", "**private_key**" or "**public_key**". Symmetric key is exploited for symmetric cryptography, while private and public keys for the public-key cryptography.

The information exchanged (also called "**data**") can be encrypted ("**ciphertext**"), or not encrypted ("**plaintext**"); the "**decryption**" Stereotype represents the process of turning ciphertext to plaintext.

During the cryptography phase the access control ("**access_Control**") consists of a set of actions that are permitted or not allowed by the system. Figure 8 shows a "**subject**" that represents an active user, a process or a device that causes information to flow among objects or changes the system state. A subject may have attributes

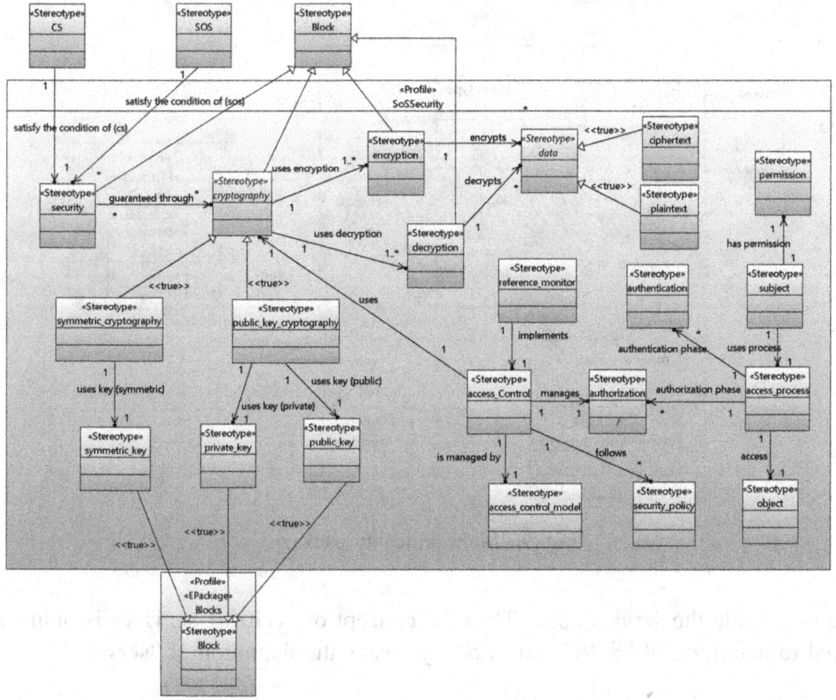

Fig. 8. SoS Security package

(“**permission**”) that describe how the subject can access to objects. An “**object**” is a passive system-related devices, files, records, tables, processes, programs, or domain containing or receiving information. Access to an object implies access to the information it contains. The “**access_process**” is composed by the “**authentication**” and the “**authorization**”. The former represents the process of verifying the identity or other attributes claimed by or assumed of a subject or verifying the source and integrity of data. The latter represents the mechanism of applying access right to a subject.

The “**reference_monitor**” represents the mechanism that implements the access control model and the “**access_control_model**” captures the set of allowed actions within a system as a policy. The access control follows a “**securityPolicy**” that represents a set of rules that are used by the system to determine whether a given subject can be permitted to gain access to a specific object.

6.3 SoS Multi-criticality Package

We introduced the concepts of “**critical_service**” as a particular type of “**service**” having a certain “**critical_level**” (see Fig. 9). The latter is associated to “**dependability_guarantee**” and “**security**”. The definition of the stereotype “**service**” belongs to the SoS Architecture package where it is linked to CS, i.e., the component, being

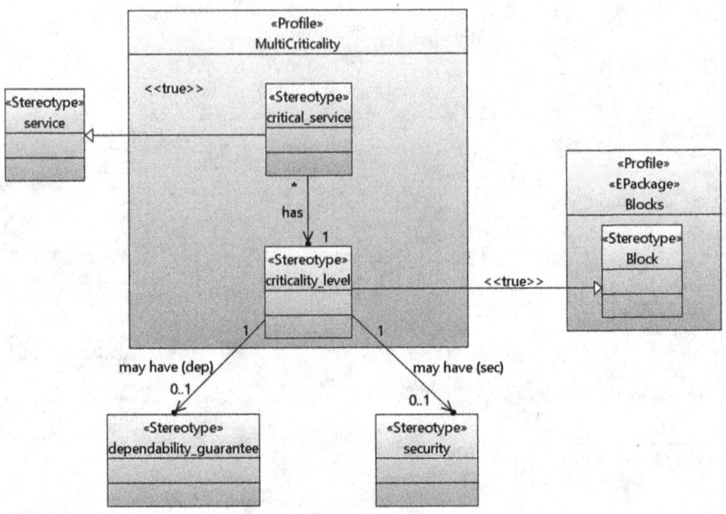

Fig. 9. Multi-criticality package

able to provide the service itself. Thus the concept of "**critical_service**" is indirectly linked to definition of "**SOS**" and "**CS**" by means the definition of "**service**".

7 Time Viewpoint

In an SoS a *global notion of time* is required in order to:

- Enable the interpretation of timestamps in the different CSs;
- Limit the validity of real-time control data;
- Synchronize input and output actions across nodes;
- Provide conflict-free resource allocation;
- Perform prompt error detection;
- Strengthen security protocols.

Time is fundamental since SoSs are sensitive to the progression of time and it is necessary to design responsive SoSs able to achieve reliably time-dependent requirements [9].

The progression of time enables change, i.e., dynamicity and evolution, in SoSs. In the AMADEOS project, it has been concluded that a *global sparse timebase* – accessible by all CSs – is fundamental for reducing cognitive complexity in understanding aspects related to all non-static investigated viewpoints on SoSs. For example, a sparse global time base allows establishing consistently – across all CSs – a temporal order among sparse events, regardless which CSs originally produced these sparse events.

We express the time-related concepts by adopting the MARTE standard [19]. MARTE is an UML profile that provides support for non-functional property modelling, defines concepts for software, hardware platform modelling, and concepts for quantitative analysis (e.g. schedulability, performance).

We measure time through clocks by defining a clock stereotype that extends the one defined in the MARTE profile. A MARTE Clock Stereotype is considered as a means to access to time, either physical or logical. The MARTE Clock is an abstract class and it refers to a discrete time.

7.1 SoS Time Package

Figure 10 shows a set of main time-related aspects. A Constituent System (defined in *SoS Architecture* package) can share a clock. The Stereotype "**clock**" is also defined as a SysML Block in order to model this concept through a Block Definition Diagram. A (digital) clock is an autonomous system that consists of an oscillator and a register. Whenever the oscillator completes a period, an event is generated that increments the register. A "**timeline**" represents the progression of the time and it is designed with a Stereotype that extends the metaclass "**Lifeline**" of a Sequence Diagram. The "**timeline**" is composed by an infinite number of instants ("**instant**" Stereotype) measured using a "**time_code**" and a "**time_scale**". A time code is a system of digital or analog symbols used in a specified format to convey time information i.e., date, time of day or time interval. A time scale is a family of time codes for a particular timeline that provide an unambiguous time ordering (temporal order of events).

A "**clock**" could be based on an "**internal_sync**", i.e., on a process of mutual synchronization of an ensemble of clocks in order to establish a global time with a bounded precision, or on an "**external_sync**", i.e., on the synchronization of a clock with an external time base such as GPS. It could be a "**reference_clock**", i.e., a hypothetical clock of a granularity smaller than any duration of interest and whose state is in agreement with TAI, or a "**primary_clock**", i.e., a clock whose rate corresponds to the adopted definition of the second (the primary clock achieves its specified accuracy independently of calibration).

Finally, the clock could have the following properties:

- "**accuracy**" - the maximum offset of a given clock from the external time reference during the time interval of interest, measured by the reference clock;
- "**granularity**" - the duration between two successive ticks of a clock; "**tick**" - the event that increments the register of the clock;
- "**offset**" - the offset of two events denotes the duration between two events and the position of the second event with respect to the first event on the timeline;
- "**frequency_offset**" - the frequency difference between a frequency value and the reference frequency value;
- "**stability**" - a measure that denotes the constancy of the oscillator frequency during the given interval of time of interest;
- "**wander**" - long-term phase variations of the significant instants of a timing signal from their ideal position on the time-line;
- "**jitter**" - short-term phase variations of the significant instants of a timing signal from their ideal position on the time-line.

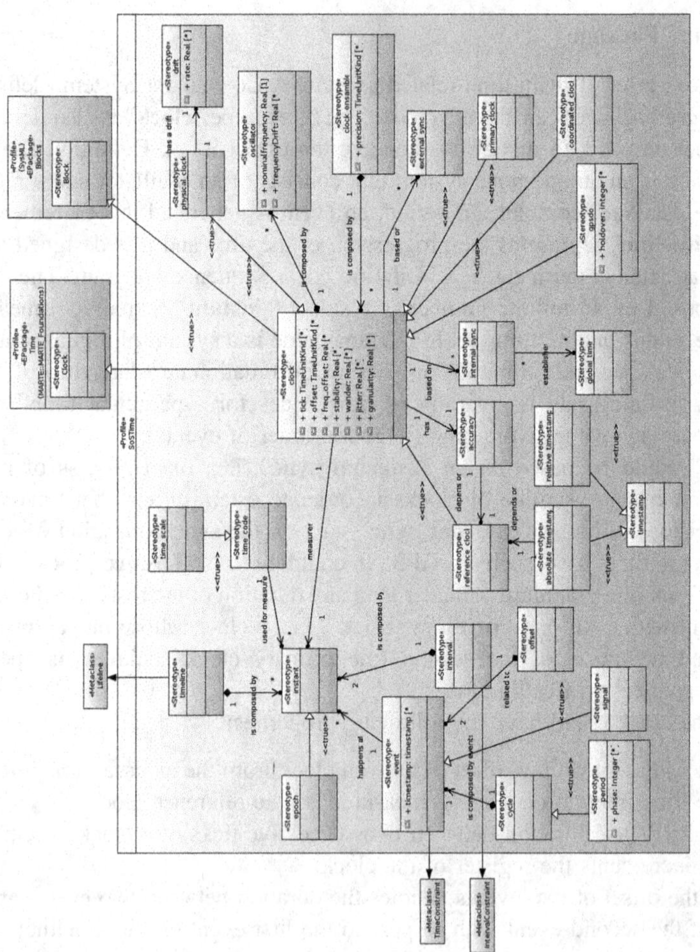

Fig. 10. SoS Time package

If a clock is a "**physical clock**", we use the "**drift**" measure in order to describe the frequency ratio between the physical and the reference clock. A digital clock consists of an "**oscillator**", represented as a Stereotype, with a "**nominal_frequency**" and a "**frequency_drift**", represented as properties. A "**coordinated_clock**" is a particular type of a clock, it is synchronized within stated limits to a reference clock. A "**clock_ensamble**" is a collection of clocks operated together in a coordinated way with a certain "**precision**". We define "**gpsdo**", a Stereotype that represents a particular type of clock where its time signals are synchronized with information received from a GPS receiver, and "**holdover**", a property expressing the duration during which the local clock can maintain the required precision of the time without any input from the GPS.

The "**timestamp**" is the state of a selected clock at the instant of event occurrence. It depends on selected clock and if we use the reference clock for time-stamping, we call the timestamp "**absolute_timestamp**". An ensemble of clocks could synchronize in order to establish a "**global_time**" with a bounded precision.

An "**instant**" is a cut of the "**timeline**" and an "**interval**" is a section of timeline composed by two instants. The latter is defined as an "**IntervalConstraint**" of a Sequence Diagram.

An "**event**" can happen at a particular instant, and to represent this type of information we have used a "**TimeConstraint**" of a Sequence Diagram. A "**signal**" is a particular event used to convey information typically by arrangement between the parties concerned. An "**epoch**" is a particular instant on the timeline chosen as the origin for the time-measurement. A "**cycle**" is a temporal sequence of significant events whereas a "**period**" is a specific type of cycle marked by a constant duration between the related states at the start and the end at the end of the cycle, called "**phase**". The offset of two events denotes the duration between two events and it is represented by the "**offset**" Stereotype.

8 Emergence Viewpoint

The concept of Emergence (see also Chap. 3 of this book) is one of the most important challenges of AMADEOS. As already described in previous sections, SoSs are built to realize new services that CSs separately cannot provide.

Emergence mainly denotes the appearance of novel phenomena at the SoS level that are not observable at CSs level; managing *emergence* is essential to avoid un-desired, possibly unexpected situations generated from CSs interactions and to realize desired emergent phenomena being usually the higher goal of an SoS [14].

In the AMADEOS conceptual model, emergence is defined as follows: '*A phenomenon of a whole at the macro-level is emergent if and only if it is new with respect to the non-relational phenomena of any of its proper parts at the micro level*'. Consequently, it is behavior observable at the global level (e.g., a traffic jam) that cannot be reduced to the behavior of one of the parts (e.g., a single car analyzed in isolation). If an emergent phenomenon can be explained by a trans-ordinal law, i.e., a law that explains the emergent phenomenon at the macro level from properties or interactions of parts at the micro level, it is *explained emergence*. In case such laws have not been found (yet), it is *unexplained emergence*. While there are cases of unexplained emergence (e.g., the human consciousness), the type of emergence that is occurring in the cyber part of an SoS is *explained emergence,* even if we are surprised and cannot explain the occurrence of an unexpected emergent phenomenon at the moment of its first encounter. If we have made proper provisions to observe and document all interactions (messages) among the CSs in the domains of time and value, we can replay and analyze the scenario after the fact. At the end, we will find the mechanisms that explain the occurrence of the emergent phenomenon. There is no ontological novelty in the interactions of the CSs in the cyber parts of an SoS.

Hence an explained emergent phenomenon can be classified as expected (trans-ordinal laws are known), or unexpected (trans-ordinal laws are not known). Orthogonally, emergent phenomenon can be *beneficial*, or *detrimental*.

Hence four cases of emergent behavior must be distinguished in an SoS. *Expected and beneficial* emergent behavior is the normal case. *Unexpected and beneficial* emergent behavior is a positive surprise. *Expected detrimental* emergent behavior can be avoided by adhering to proper design rules. The problematic case is *unexpected detrimental emergent behaviour*. For an in-depth discussion about emergence in SoSs we refer to [12].

8.1 SoS Emergence Package

In this section we show how to use a semi-formal language to represent an emergent behavior of an SoS. Nevertheless, because of the nature of the emergence concept, defining a semi-formal language, thus only eliciting an emergent behavior, is not sufficient. Our aim is also capturing operational aspects related to emergence by considering an SoS in action.

For these reasons we propose two different types of representation that a system designer can choose:

- Block Definition Diagram;
- Sequence Diagram.

Figure 11 shows the profile package for the emergence behavior as a block definition diagram.

This package represents the main concepts of emergence using a Block Definition Diagram. We represent a "**phenomenon**" as a block and we distinguish an "**emergent_phenomenon**" from a "**resultant_phenomenon**". An emergent phenomenon can be explained ("**explained_emergence_phenomenon**") or unexplained ("**unexplained_emergence_phenomenon**") and in the former case there is a trans-ordinal law ("**transOrdinal_law**") that explains the behavior.

An SoS with emergent phenomena has an emergent behavior that could be expected, unexpected, beneficial or detrimental. For this reason, we consequently defined the four following blocks: "**unexpected_and_detrimental**", "**expected_and_detrimental**", "**unexpected_and_beneficial**", "**expected_and_beneficial**".

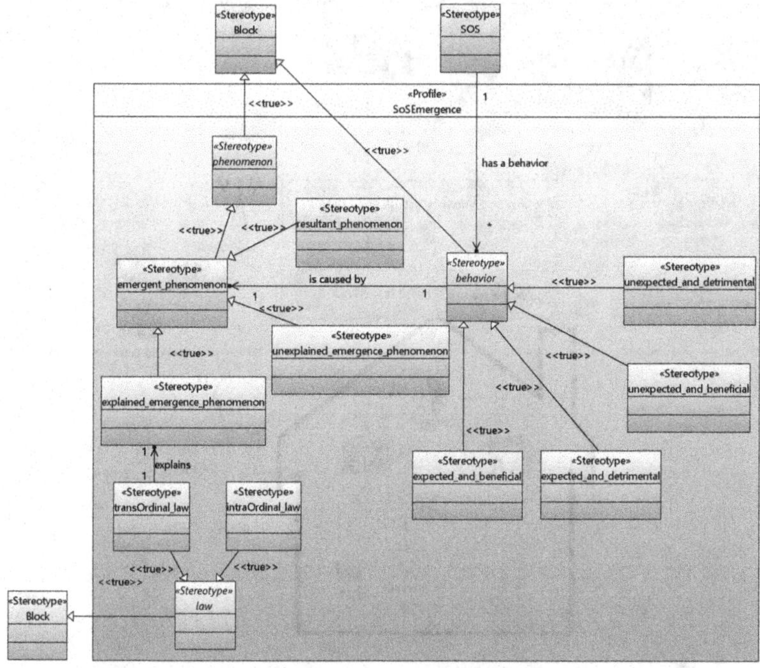

Fig. 11. SoS Emergence package

9 The Profile at Work

In this section we introduce a Smart Grid household scenario to exemplify the application of the profile and to instantiate the basic SoS concepts to a concrete case-study from the Smart Grid domain, focusing on the Architecture (Sect. 9.1) and Emergence (Sect. 9.2) viewpoints. Further examples of application of the profile to the selected use-case can be found in [15] and public deliverable D2.3 "AMADEOS conceptual model - Revised" [1].

In a Smart Grid household scenario different operationally independent subsystems aim at delivering the desired emergent phenomenon of improving the efficiency and the reliability of the production and distribution of electricity through communication facilities. Requests for energy coming from *electronic appliances* are forwarded towards the subsystems in charge of granting or denying each request while achieving the Smart Grid goal, i.e., keeping the production and consumption rates for connected households balanced.

Figure 12 shows the topology of the main subsystems involved within a single household of the Smart Grid scenario. Washing machines and microwaves are examples of electronic appliances. They represent a *flexible load* which may initiate an energy request. The *smart meter* measures energy consumption and production rates; the *Distributed Energy Resource* (*DER*) manages the energy produced through energy generating and storage systems, like wind-powered electrical generators or batteries.

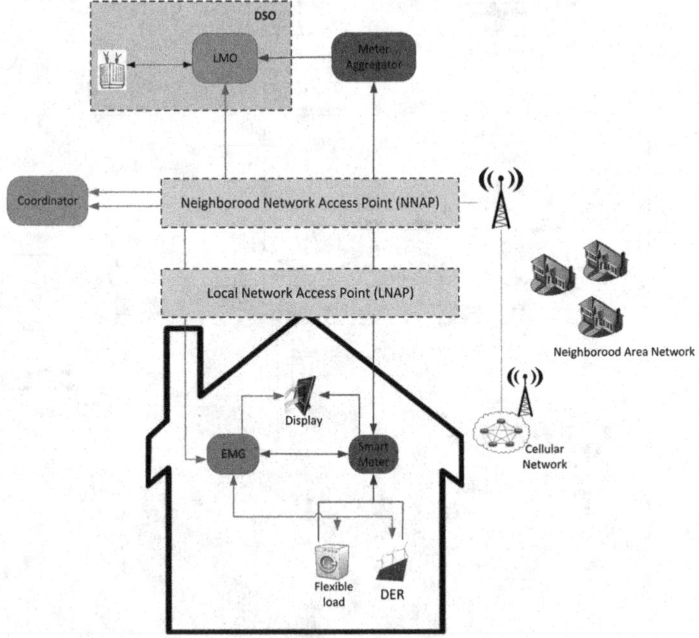

Fig. 12. Smart Grid case study

A *command display* shows consumption rates and enables residents to interact with their own energy control system. The *Energy Management Gateway* (*EMG*) controls the flexible loads and the DER based on measurements received from the smart meter and in agreement with the *coordinator* to establish optimal energy distribution. The coordinator is connected to the *Neighborhood Network Access Point* (*NNAP*) with the aim of keeping the production and the consumption of energy for a set of connected households balanced. A *Distribution System Operator* (*DSO*) regulates consumption and production rates at the country level. By means of its *Load Management Optimizer* (*LMO*), a DSO receives information from a *meter aggregator* and enacts control decisions in cooperation with the coordinator. The access to the household is provided by one or more *Local Network Access Points* (*LNAPs*) connected to a NNAP. All the above mentioned components require proper interfaces in order to exchange control messages and physical energy entities within and outside the household Smart Grid.

9.1 Modeling the Architecture Viewpoint

Using the SoS Architecture package it is possible to represent the topology of any System of Systems. Now we show how to use SoS through the Smart Grid household case study.

First of all, it is necessary to decide what are the main constituent systems involved, and how to represent them. For each system component we use a Block element of a Block Definition Diagram and through the connections we show the relations between

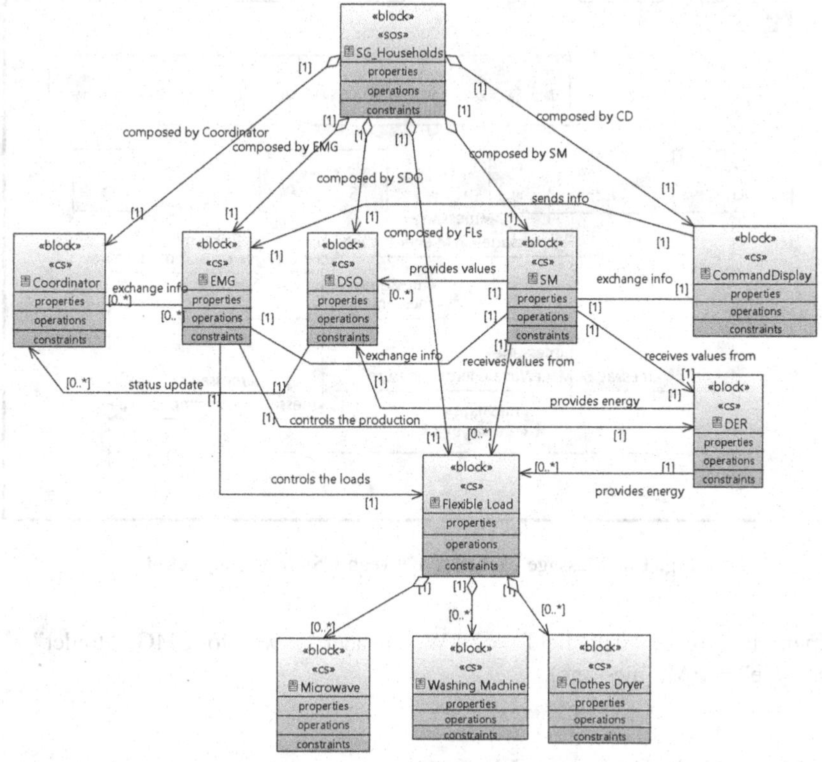

Fig. 13. Smart Grid household Block Definition Diagram

them. Using the stereotypes defined in the "*SoS Architecture*" package it is possible define the Smart Grid household as a system of systems (SoS) and all the other elements as constituent systems (CSs).

Figure 13 shows a model example of a Smart Grid with the application of our profile ("*SoS Architecture*" package). The "SG_Households" is a Block and it is stereotyped as an SoS; it is composed by 5 CSs, which exchange information. Among others, the block "Flexible Load" is stereotyped as a CS and it is composed by a set of household electrical appliances: Microwave, Washing Machine, Clothes Dryer, etc. These latter are switched on and off dynamically based on the current needs.

An application example of the main SoS communication concepts is shown in Fig. 14. Through the Smart Grid household case study, we describe a set of communication messages exchanged between the involved CSs.

First of all, it is necessary to decide which are the involved elements in the communication and how many message are exchanged. We identify a "Lifeline" as a constituent system and a "message" as exchange data between two constituent systems. A message could contain all the properties defined in Fig. 2 and they can be displayed using a constraint or a comment box. Figure 14 shows the message properties using a

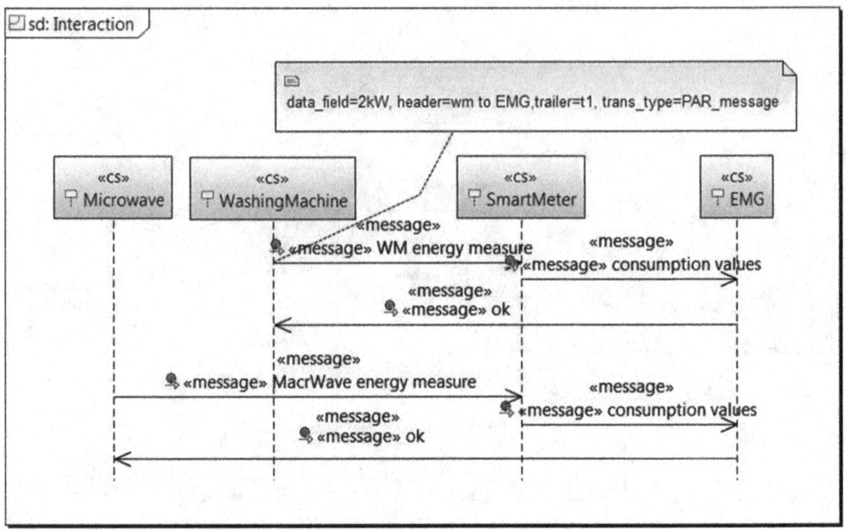

Fig. 14. Message exchange between CSs of a Smart Grid

comment box (e.g., "data_field" = 2 kW, "header" = wm to EMG, "trailer" = t1, "trans_type" = PAR_message).

9.2 Modeling the Emergence Viewpoint

While a Block definition diagram defines stereotypes and related elements to capture statically the emergence behavior, a sequence diagram is able to define dynamic interactions leading to emergence. We adopt a sequence diagram where each lifeline represents a constituent system and each message specifies the kind of communication between the lifelines, the sender and the receiver. An SoS is prone to changes: sometimes constituent systems are incremented, modified or removed. To this end, this kind of diagram helps the system designer to easily update and analyze new system behaviors. The diagram not only describes the communication but it also helps to represent the SoS behavior during the progression of time.

To show the difference between a "static" and "dynamic" representation of emergence, we consider a particular scenario of the Smart Grid previously described. The dynamicity of the household electrical appliances may lead to an emergent behavior of the system in case of a peak of request of energy coming from the neighborhood. Let us assume that because of a public event, an exceptional lighting of specific public spaces has to be supported by the Smart Grid. In this case, while in the household it was commonly possible to turn on microwave and washing machine together, we end up in a very limited provision of energy, which is not sufficient for both the electrical appliances. This phenomenon represents an *emergent behavior* of the Smart Grid since it is not possible to devise it if we only look at the interactions of the internal household CSs without considering the neighborhood CSs.

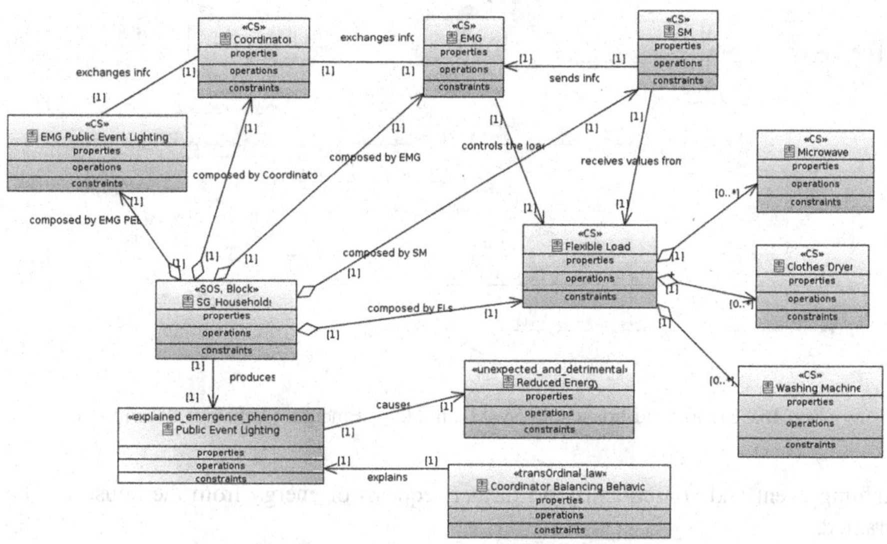

Fig. 15. Smart Grid household – Emergent behavior

Figure 15 shows, through a BDD, how the public event lighting is represented as an explained and detrimental emergent phenomenon, explained by the balancing behavior of the Coordinator and causing reduced energy for the electrical appliances. This phenomenon causes an *unexpected* and *detrimental* behavior of the SoS, which allows it to only satisfy a subset of energy requests.

However, using this type of diagram we are not able to represent the progression of time and the semantic of message that may contribute to reveal emergence phenomena. Especially, the above representation does not attach greater importance to capture the time aspects of the SoS emergent phenomena.

We now introduce the representation of the exceeding peak energy request by using a sequence diagram. We adopt a sequence diagram to show the emergent behavior of the electrical appliances request by means of the interaction among related CSs of the Smart Grid.

As shown in Fig. 16, "electronic appliances" CSs are represented as "Lifeline" and their interactions are represented through directed labeled arrows. Washing Machine is switched on at t2 after the agreement allowed from the Coordinator. As next, the Coordinator receives (at time t3) and grants (at time t4) the energy for switching on the public lighting for the exceptional event. This request is forwarded to the Coordinator by the Public Event Lighting (PEL) EMG, which is external to the household. At time t5, microwave issues its request to be connected to the Smart Grid but it receives a negative acknowledgment at time t7. Usually, the household would be able to switch on the washing machine and the microwave at the same time. On the contrary, because of the public event lighting resulting in a peak of energy from the house neighborhood it results that only a reduced amount of energy is available for the electrical appliance (emergent behavior). Indeed, right before the requests issued from the microwave (time t5) and the clothes dryer (time t8), the Coordinator allocates the energy for the public

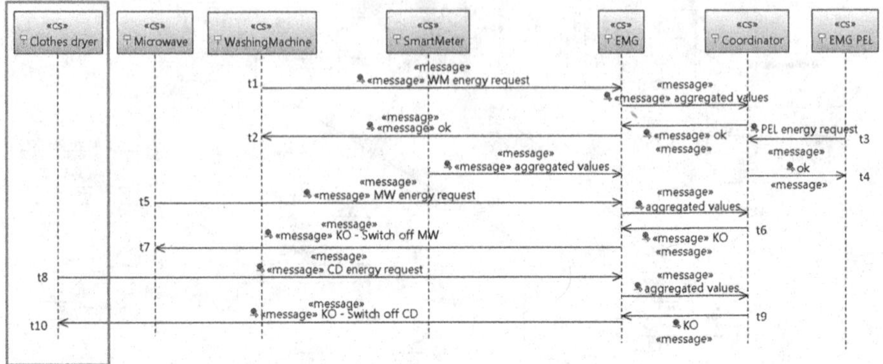

Fig. 16. Smart Grid household SysML model – Emergent behavior description

lighting event and consequently no further requests of energy from the house can be granted.

This illustrative example shows that networked individual systems together to realize a higher goal, which none of individual system can achieve in isolation, could lead to an emergent behavior: impossibility of satisfying commonly granted energy requests. The Emergent behavior is shown through the message exchange and it consists of unexpected and detrimental emergent behavior caused by a system dynamicity property.

10 Related Works

In this section we present an overview of related ADL design approaches presented in the literature of SoSs. This analysis is not meant to be exhaustive but it is based on some of the most representative related works on designing SoSs. Its objective is to determine to what extent viewpoints-based SoS concepts have been already captured in the literature.

In [7] the authors propose the use of SysML in representing an SoS by adopting and in some cases extending canonical SysML diagrams in order to model different viewpoints of an SoS. Beyond *structure*, a specific support to the *multi-criticality* viewpoint is also provided by adopting the specific stereotypes aiming at grouping requirements according to qualitative and quantities metrics to support trade-off analysis. Nevertheless, there is no specific support for other viewpoints, including *time*, *dependability/security*, *dynamicity*, *evolution* and *emergence*.

A partial answer to the above issues is given by the approach presented in [13] providing support to *structure* and *evolution* viewpoints of an SoS by exploiting several SysML models. The authors propose the adoption of diagrams to determine an evolving SoS and its environment and the interactions occurring between an SoS and the environment and among CSs themselves. Noteworthy the approach is still missing specific support to *dynamicity*, *emergence*, *multi-criticality*, *dependability/security* and *time*. In [20], the presented SysML modeling approach allows the definition of the SoS *structure* and how to support *dynamicity* and *evolution* viewpoints by means of understanding the dis-alignment of a simulated SoS with respect to its requirements. Noteworthy, it is still missing a specific support to *emergence*, *multi-criticality*, *dependability*, *security* and *time*.

The approaches presented in [3, 8] provide support to model the *structure* of an SoS and *emergence* by means of the extension to SysML diagrams. Analyses of the former models are conducted to provide evidence that requirements are fulfilled. The approach supports fault-handling (*dependability* viewpoint) and responsiveness (*time* viewpoint) of an SoS, but it does not provide any specific support to *dynamicity*, *evolution* and *multi-criticality*.

The approach in [6], within the context of the DANSE EU project [4], supports the definition of an SoS *structure*, *dynamicity* and *evolution* (by means of Graph Grammars), *emergence*, etc., with the only exception of *multi-criticality*. DANSE presented a set of methodologies and tools to model and to analyze SoSs based on the Unified Profile for DoDAF and MoDAF (UPDM). In particular, DANSE focuses on the six models that can be represented as executable forms of SysML as partially reported in [6], according to a well-defined formalism to relate basics SoS concepts and their relationships. In the context of DANSE, the Goal Specification Contract Language (GSCL) assures the achievement of dependability and security requirements and it guarantees the timely response of an SoS.

All these approaches have shown the utility of adopting SysML formalisms to model architectural aspects of SoSs, thus supporting different types of analysis and a first step towards executable artifacts which can be automatically derived. Although these approaches provide detailed insights for different viewpoints aspects, it is still missing (i) an homogeneous synthesis at a more abstract level of key design-related SoS concepts, and (ii) a viewpoint-based vision. Bringing this perspective in one single consistent reference model, it is possible to provide solutions to specific design problems while still keeping the required interconnections among viewpoints.

11 Conclusions

This chapter presented a viewpoint-driven approach to design SoSs by adopting a SysML profile. We pointed out the gaps in the literature of ADLs for SoSs with respect to a set of viewpoints that we deemed essential for understanding SoSs. We outlined the conceptual model at the basis of the profile and we presented how to solve specific viewpoint needs in an integrated fashion by exploiting the high-level SoS representation in a small scale scenario. We implemented the profile in the Eclipse integrated development environment jointly with Papyrus [5], i.e., an Eclipse plug-in supporting advanced facilities for manipulating UML artifacts and SysML profiling.

The AMADEOS SoS profile can be adopted along with a Model-Driven Engineering (MDE) approach. MDE is an approach to system development and it provides a means for using models to direct the course of understanding, design, construction, deployment, operation, maintenance and modification. For a model-driven architecture perspective [18], our SoS profile is a Platform Independent Model (PIM) or, in other words, a view of the system from the platform independent viewpoint. It provides a set of technical concepts involving SoS architecture and behavior without losing the platform independent characteristics.

This kind of independent architecture makes possible to analyze step by step all the PIM viewpoints and to obtain one or more Platform Specific Models (PSM), where the SoS profile is specialized and improved according to the domain/enterprise specific technologies that belong to the enterprise implementing the SoS instance. A PSM is a view of a system from the platform-specific viewpoint. It combines the specifications in the PIM with details that specify how that system uses a particular type of platform and on the platform itself.

Furthermore, the SoS PSM can represent the base step for other activities such as the following:

- Source code generation: through an automatic transformation the SoS model can be translated in source code;
- System analysis: the SoS model can be the starting point for a lot of system analysis like: hazard analysis (HA), Failure Mode and Effect Analysis (FMEA), Fault Tree Analysis (FTA);
- System testing: the SoS model can be the basic layer to identify test procedures or resolve problems of testing coverage.

References

1. AMADEOS project - Public Deliverables. AMADEOS project. http://amadeos-project.eu/documents/public-deliverables/. Accessed 28 Sep 2016
2. Avizienis, A., Laprie, J., Randell, B., Landwehr, C.: Basic concepts and taxonomy of dependable and secure computing. IEEE TDSC **1**(1), 11–33 (2003)
3. Bryans, J., Fitzgerald, J.S., Payne, R., Kristensen, K.: Maintaining emergence in systems of systems integration: a contractual approach using SysML. In: INCOSE (2014)
4. DANSE: DANSE Methodology V2 - D_4.3 (s.d.). https://www.danse-ip.eu
5. ECLIPSE - MDT/Papyrus: Eclipse Model Development Tools (MDT) (n.d.). http://wiki.eclipse.org/MDT/Papyrus-Proposal. Retrieved 14–29 Sep 2016
6. Gezgin, T., Etzien, C., Henkler, S., Rettberg, A.: Towards a rigorous modeling for-malism for systems of systems. In: ISORCW, pp. 204–211. IEEE (2012)
7. Huynh, T.V., Osmundson, J.S.: An integrated systems engineering methodology for analyzing systems of systems architectures. In: Asia-Pacific Systems Engineering Conference, Singapore (2007)
8. Ingram, C., Fitzgerald, J., Holt, J., Plat, N.: Integrating an upgraded constituent system in a system of systems: a SysML case study. In: INCOSE International Symposium (2015)
9. Kopetz, H.: Real-time Systems: Design Principles for Distributed Embedded Applications. Springer, New York (2011)
10. Kopetz, H.: Conceptual model for the information transfer in systems of systems. In: ISORC, pp. 17–24. IEEE Press (2014)
11. Kopetz, H., Fromel, B.: Direct versus stigmergic information flow in systems-of-systems. In: System of Systems Engineering Conference (SoSE). IEEE (2015)

12. Kopetz, H., Höftberger, O., Frömel, B., Brancati, F., Bondavalli, A.: Towards an understanding of emergence in systems-of-systems, pp. 214–219. IEEE
13. Lane, J.A., Bohn, T.: Using SysML modeling to understand and evolve systems of systems. Syst. Eng. **16**(1), 87–98 (2013)
14. Mogul, J.: Emergent (Mis)behavior vs. complex software systems. In: EuroSys, pp. 293–304. ACM (2006)
15. Mori, M., Ceccarelli, A., Lollini, P., Bondavalli, A., Frömel, B.: A holistic viewpoint-based SysML profile to design systems-of-systems. In: International Symposium on High Assurance Systems Engineering (HASE 2016), pp. 276–283. IEEE, Orlando (2016)
16. Murer, S., Bonati, B., Furrer, F.J.: Managed Evolution: A Strategy for Very Large Information Systems. Springer, New York (2010)
17. Nakagawa, E.Y., Gonçalves, M., Guessi, M., Oliveira, L., Oquendo, F.: The state of the art and future perspectives in Systems-of-Systems software architectures. In: SESoS, pp. 13–20 (2013)
18. OMG: MDA Guide Revision 2.0, 18 June 2014. http://www.omg.org/cgi-bin/doc?ormsc/14-06-01.pdf. Accessed 14 Sep 2016
19. OMG: A UML Profile for MARTE: Modeling and Analysis of Real-Time Embedded systems, Beta 2. Document Number: ptc/2008-06-0 (s.d.)
20. Rao, M., Ramakrishnan, S., Dagli, C.: Modeling and simulation of net centric system of systems using systems modeling language and colored petri-nets: a demonstration using the global earth observation system of systems. Syst. Eng. **11**(3) (s.d.)
21. Schmerl, B., Aldrich, J., Garlan, D., Kazman, R., Yan, H.: Discovering architectures from running systems. IEEE TSE **32**(7), 454–466 (2006)
22. Selberg, S.A., Austin, M.A.: Toward an evolutionary system-of-systems architecture. In: INCOSE, pp. 1065–1078 (2008)
23. Verissimo, P.: Travelling through wormholes: a new look at distributed systems models. SIGACT News **37**(1), 66–81 (2006)

AMADEOS Framework and Supporting Tools

Arun Babu[1], Sorin Iacob[2], Paolo Lollini[3(✉)], and Marco Mori[3]

[1] Resiltech SRL, Pisa, Italy
arun.babu@resiltech.com
[2] Thales Nederland B.V., Hengelo, The Netherlands
sorin.iacob@nl.thalesgroup.com
[3] Department of Mathematics and Informatics,
University of Florence, Florence, Italy
{paolo.lollini,marco.mori}@unifi.it

1 Introduction

This chapter defines the overall tool-supported "AMADEOS architectural framework", with its main building blocks and interfaces. It particularly focuses on Structure, Dependability, Security, Emergence, and Multi-criticality viewpoints of an SoS. Finally, for SoS modeling, a "supporting facility tool" based on Blockly is demonstrated. Blockly is a visual DSL and has been adopted to ease the design of SoS by means of simpler and intuitive user interface; thus requiring minimal technology expertise and support for the SoS designer.

2 Architecture Framework for SoS

Architectural framework does not refer to the specific design of specific system architecture, but they rather represent a view on how such architecture should be described. Although architectural frameworks are "prescriptive" and not "descriptive", there is still no consensus on providing a methodological step-by-step instruction to be followed. In [1], the authors describe a study involving the use of a design approach to guide the development of an SoS architecture by means of rules, guidance and artefacts for collaboratively developing, presenting and communicating architectures without an order set of phases to carry out. In [2, 3], it is noticed a close relation between architecting methods and the architectural frameworks, thus a step-by-step set of instructions is provided to guide the development of SoS architectures.

When building an SoS architectural framework, the aim is to be instrumental in the creation of future evolvable systems of systems. Both description views and methodology shall be allowed, as long as it facilitates the design of the architecture of such systems.

Architectural frameworks that are currently used in SoS literature have been applied in different contexts of operation along with ADL solutions to model different

This work has been partially supported by the FP7-610535-AMADEOS project.

A. Bondavalli et al. (Eds.): Cyber-Physical Systems of Systems, LNCS 10099, pp. 128–164, 2016.
DOI: 10.1007/978-3-319-47590-5_5

architectural aspects of an SoS. In the following we provide insights on currently adopted architectural frameworks and ADL approaches.

2.1 ADLs in SoS Architectural Frameworks

This section collects a few ADL approaches that have been proposed in the literature to model different aspects of SoS. They range from approaches dealing with very specific problems to frameworks.

Among the approaches presented in the context of research projects we consider solutions proposed in COMPASS and DANSE EU projects. COMPASS aims at supporting the application of formal analysis tools and techniques at the level of the graphical notations used in current industrial practice. COMPASS project exploits the Artisan Studio tool [4] in order to support system and requirements modelling using SysML as well as software modelling using UML and code generation. As stated in [5] COMPASS proposes the adoption of *Context Diagrams*, *Use Case Diagrams*, *Block Definition Diagrams* and *Sequence Diagrams*. COMPASS exploits tool's well-established extension mechanisms to extend traditional systems modelling as needed to model SoS. Starting from artefact created with the tool, COMPASS provide a well-defined denotational semantic of SysML blocks by means of the COMPASS modelling language (CML), a formal specification language that supports a variety of analysis techniques.

The DANSE methodology and tools are mainly based on the Unified Profile for DoDAF and MoDAF (UPDM). The latter has also been extended to cover the NATO Architecture Framework (NAF) and it provides more than fifty different model types grouped in eight viewpoints [6]. These viewpoints are: *Capability Viewpoint*, *Operational Viewpoints*, *Service Viewpoint*, *System Viewpoints*, *Service Viewpoint*, *Data & Information Viewpoint*, *Project Viewpoint* and *Standard Viewpoint*. In particular DANSE focuses on the six models that can be represented as executable forms of System Modelling Language (SysML).

In [7], the authors propose a formalism for relating basics SoS concepts by means of a UML class diagram. They identify as basic concepts *SystemType*, *System-Of-Systems*, *Goal*, *Role*, *Service*, *Requirement*, *Port*, *Requirement* and *Port*. Consequently, they adopted their defined formalism to instantiate an operative SoS by means of adopting canonical UML diagrams such as Sequence diagram. The behaviour of CS is formalized through Timed Automata and its dynamicity/evolution is achieved by means of Graph Grammars.

An example of modelling SoS by means of SysML is given in [8] where the authors exploit different diagrams and in particular executable diagram in order to simulate Net-centric SoS through the Petri Net formalism. In [9] the authors propose the use of SysML in representing an SoS in general and for a particular applicative scenario. They propose to adopt and in some cases to extend canonical SysML diagrams in order to model different aspects of an SoS. They defined concept Diagram as an extension of class diagrams to depict the top-level systems of an SoS and external stereotypes. This helps in identifying the boundaries between the system and its environment. They adopted the *class diagram* with an aggregator operator to represent that a component is composed by a set of other components. They proposed the adoption of a *requirement diagram* with an additional stereotype, i.e., *critical requirement* which is a particular type of *requirement*. This diagram groups together requirements according to qualitative and quantities metrics to support a trade-off analysis. They adopt canonical *use case diagrams* to represent the set of action an SoS performs. The SysML *activity* and *sequence diagram* are exploited to represent the SoS at the functional level and its exchanges of messages, respectively. Finally, they exploit a *block diagram* as a refinement of their *concept diagram,* which aims at representing blocks/component with well-defined interfaces, i.e., *serviceports* and *flowports*.

The approach presented in [10] describes how several SysML models can be used to support a set of needs that the authors deemed essential for an SoS, namely *translating capability objectives*, *understanding systems and their relationships*, *monitoring* and *assessing changes*, *developing and evolving the SoS architecture*, *addressing requirements* and *solution options*. The authors propose to apply a Model-Driven Systems Development (MDSD) approach [11] to an SoS. The first step consists in determining capabilities and actors through *use cases diagrams* by defining what is in the system and what remains outside, as stated in a *context diagram*. Use cases determine the top-level service or capabilities and the major actions necessary to perform the use cases and all of the alternate actions. Finally, two different diagrams describe the interactions, i.e., *black box sequence diagram* and *white box sequence diagram.* Black box sequence diagrams show the flow of the requests that pass between the SoS and the environment while white box sequence diagrams depict the flow of requests between the constituent systems, and between the constituent systems and the external entities.

Among others, the approaches presented in this section show the utility of adopting SysML formalisms in order to model different architectural and non-architectural aspects of an SoS. This supports different types of analysis and it represents a first step towards executable artefacts, which can be automatically derived from SysML. As shown in this section, in the literature different attempts exist to apply SysML approaches to specific viewpoints that we deemed essential in providing architecture for Multi-Critical Agile Dependable and Evolutionary SoS. Nevertheless, an architectural framework that provides an integrated support to all these viewpoints is still missing. The architectural framework will benefit of the approaches proposed in the literature in supporting specific viewpoints (when they exist) and it will integrate SysML specific solutions to provide a usable high-level support for designers of SoS.

3 The AMADEOS Architecture Framework

The AMADEOS architectural framework (AF) is described by means of a high-level perspective of activities and artefacts involved in SoS design phases and by its viewpoint-based specialization.

3.1 High-Level View

The high-level representation of the AF is shown in Fig. 1 as a pyramid made of four different layers, namely *Mission, Conceptual, Logical* and *Implementation.* Apart from the *Mission* block, all the remaining levels are organized in slices, each corresponding to a specific viewpoint.

The starting point of the AF consists in defining the ***Mission*** of an SoS. The mission is commonly formalized by means of a document of intents created by enterprise managers having in mind a high-level perspective of the system and a clear definition of business-related issues. The document of intents is written in natural language to formalize the overall objectives and functionalities of an SoS starting from a shortened version of the glossary illustrating main SoS concepts and other related mission-relevant arguments.

At the ***Conceptual Level***, it is possible to consider a subset of viewpoints depending on the target SoS and its mission; however in AMADEOS we focus till to collaborative SoS, for which we identify a set of viewpoints that must be considered as mandatory. Inputs to these levels are the document of intents describing the mission, the conceptual model [12] defining main SoS concepts and their relationships and the AMADEOS meta-requirements [13], which can guide the identification of requirements for specific SoS instances. For each viewpoint (corresponding to a slice of the

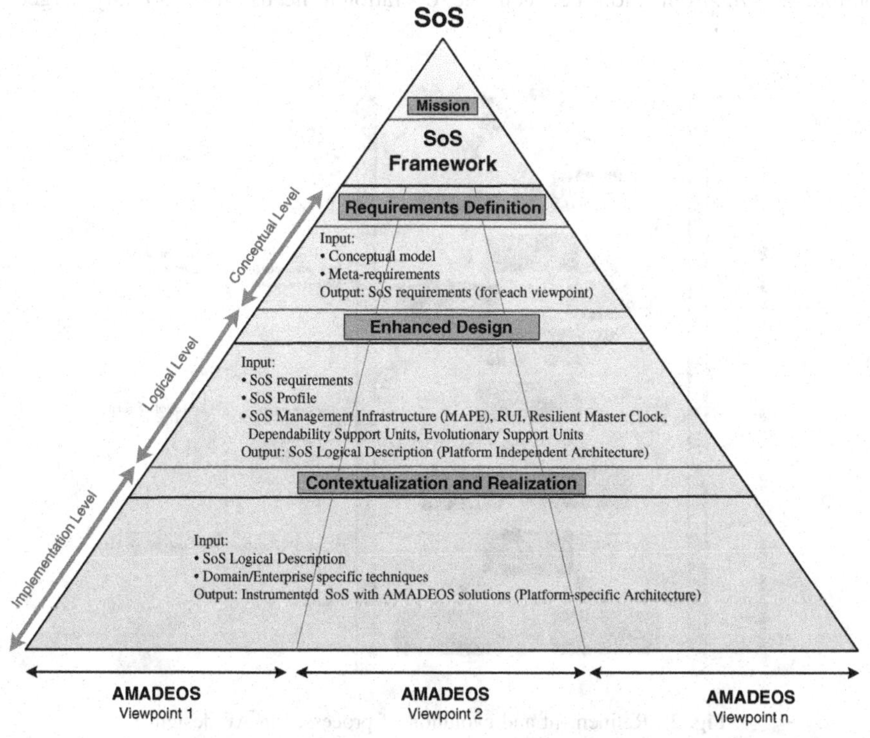

Fig. 1. AMADEOS architectural framework

pyramid), the SoS is examined and described. The resulting description should be the requirements of the SoS (these can be expressed in natural language, as well as using formalisms for the description of requirements). The identification of relations between the different viewpoints is carried out at this phase.

The ***Logical Level*** provides support for designing an SoS based on the viewpoints requirements in the AMADEOS SysML profileand the Building blocks defined in Sect. 4. The output of this phase consists in the platform independent description of the SoS in a semi-formal language (SysML), for the different viewpoints.

The ***Implementation Level*** leads to the integration of new CSs with already existing and deployed CSs starting from the logical architecture defined at the previous level and domain/enterprise specific techniques. At this level, the input logical architecture is refined and instrumented with domain/enterprise specific technologies which belong to the enterprise implementing the SoS instance.

We depict in Fig. 2 a process-based view of each level of the AF. We represent the basic task ad the input/output artefacts involved at each level. This gives a more detailed description of the relationships and evolution of the main artefacts produced in each level of the pyramid (see Fig. 1) and the relations between levels (through the top-down processes of refinement and instantiation, and bottom-up processes of generalization and abstraction). The artefacts categories at each level are intended to be generic enough to fit all the viewpoints.

On the *Mission Level*, a relatively slow-paced cycle takes place to address the continuous synchronization between the operational needs, the currently targeted

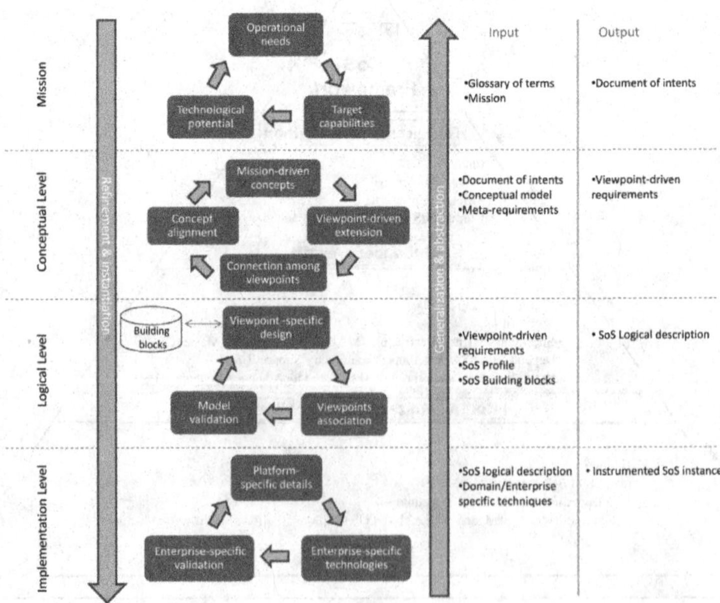

Fig. 2. Refinement and evolution of processes in AF design

capabilities of the SoS architecture and the technological possibilities to achieve the needs. At this stage, enterprise managers iterate the above phases to determine the mission of the SoS which is then formalized in a document of intents along with possible target solutions to be implemented.

On the *Conceptual Level* the alignment between the overall envisioned concepts and the SoS domain takes place more frequently. On this level the AMADEOS concepts which are relevant to achieve the mission are extracted from the document of intents and then filtered based on the viewpoint to which they belong. Further details may be added at each viewpoint descriptions to support the targeted capabilities within the SoS domain. Connections among viewpoints descriptions are identified in an early stage before similar concepts are aligned with each other, if needed.

On the *Logical Level*, cycles occur at a more rapid pace and are used to ensure that all desired functionalities and qualities are supported by the developed architecture. To this end, building blocks are selected and further integrated to obtain a design model which is generic enough to be applied to different types of platform. The design process follows a viewpoint-based perspective based on which target models are created for each viewpoint. Application specific details are added at this stage before viewpoint models are linked with each other according to the dependencies early identified at the *Conceptual level*. At *Logical level*, wrappers for legacy CSs have to be defined in terms of proper RUI interfaces which connect such legacy components to the rest of the logical SoS. Finally, validation activities take place, e.g., either by supporting the generation of models that are correct by construction or through predefined consistency checks. The generation of models, the integration of building blocks and the model consistency checks are made possible by exploiting the AMADEOS profile.

The most frequent cycles occur at the *Implementation Level*, where the SoS architecture is defined at its most fine-grained level by augmenting it with specific platform-dependent and specific technologies which are exploited by the target enterprise in order to obtain an operational SoS instance. At this stage the possibly available legacy components may be added to the platform-dependent architecture, provided that they have been properly encapsulated in the SoS at the logical level. This implementation model may then be validated through the technologies which are commonly adopted in place by the enterprise. In order for this phase to be supported, it is necessary that specific validation techniques adopted in the enterprise comply with the AMADEOS profile specification. However, it is not the main focus of AMADEOS to provide full support to implementation of single CSs. Nevertheless, this phase includes all the steps from the platform independent architecture to the architecture showing how each and every feature in the product should work and how every component should work. This phase is kept in the framework for completeness.

3.2 Viewpoint-Driven Analysis

The AF has been represented through a high level view which describes the processes of defining an operational SoS instance starting from the mission definition. The architectural viewpoints required for supporting this definition are the ones considered in the AMADEOS vision, i.e., structure, dynamicity, evolution, dependability and security, time, multi-criticality, emergence. We describe in the following how the AF can support the activities required by each viewpoint.

Viewpoint of Structure. The Structure viewpoint concerns with representing the overall structure of the SoS. It focuses on architectural concerns of an SoS and it is closely related to other issues like SoS constraints, RUMI and semantics of communication. Indeed, defining interfaces among CSs is important as this stage to support their communications.

The input to the conceptual level of an SoS is the mission (or vision). This entails the overall objectives of the SoS as well as the required functionalities. The structure viewpoint entails examining these objectives and determining the constraints of the interfaces, and the communication, between constituent systems. Unlike the other viewpoints, the structure viewpoint places restrictions upon the activities of the SoS.

By addressing the meta-requirements in the context of the specific mission, a set of structural requirements can be identified that restrict the overall architecture that will be eventually delivered. For example, from [CONSTR 11], standards compliance of one or more CSs may be very important, particularly in use cases such as in the Smart Energy domain. Also the conceptual model is exploited as the vocabulary of concepts to be adopted.

The output of the conceptual level consists of a set of requirements that relate to the structural architecture of the SoS.

The logical design of an SoS architecture will be based upon the Structure requirements identified at the conceptual level and the building blocks identified along with the SysML profile. The SysML Block Definition Diagram (BDD) is used to model the topology and the relations of an SoS. Blocks in SysML BDD are the basic structural element used to model the structure of systems. A Block provides a unifying concept to describe the structure of an element or a system. This type of diagram helps a system designer to depict the static structure of an SoS in terms of its CSs systems and their possible relationships. By means of BDDs it will be possible to model the static structure of CSs, their interfaces and how the communication among CSs is achieved.

The output of the logical level will be a platform independent SoS architecture specification from a structural point of view. This will consist of the outline of the CSs identified by the requirements and the RUMIs that specify the interactions between these former CSs.

On the implementation level, the platform independent structural design from the enhanced design level is further concretized using specific contextual requirements, towards building the SoS structural architecture. For example, specific CSs may already exist and may need to be integrated. In the structural viewpoint, this will lead to specific RUMIs that are used to define how CSs will interact. These RUMIs consist of the communication protocols that define the messages that will be shared between CSs. The implementation level is very specific to the actual CSs involved and the operational context. The output consists in a fully contextualized SoS structural architecture.

Viewpoint of Dynamicity. Dynamicity refers to short-term changes in an SoS, which occur in response to changing environmental or operational parameters of the CSs. These changes can refer to offered services, built-in structure and interactions with other entities, and may have different effects, such as SoS adaptation or the generation of emergent phenomena.

Starting from the SoS mission, the dynamicity requirements and the conceptual model, the output of the conceptual level is the set of dynamicity requirements, i.e., requirements related to the dynamicity viewpoint for the specific mission. The latter are the input of the logical level which also exploit the SysML profile and the building blocks to support the generate of the platform independent SoS architecture. The building blocks of the SoS management infrastructure defined are exploited to achieve dynamicity requirements, through the monitoring, analysis, planning and execution activities. Instantiation of the profile is connected with the Structure viewpoint of the SoS. Interactions elicited among CS take into account the service provided at the RUI interfaces as regulated by the SLA.

At the implementation level, the generic SoS architecture is instantiated into a platform-specific SoS architecture. This includes, among others, the implementation of RUIs that integrate monitoring and execution features that implement the MAPE-K architecture, and SLA-oriented reconfiguration operations. It results an architecture specialized by the enterprises with their adopted technologies to provide support to dynamicity though a platform-specific architecture.

Viewpoint of Evolution. Large scale Systems-of-Systems tend to be designed for a long period of usage during which the demands and constraints put on the system will usually change, as well as its environment. Evolution is the process of gradual and progressive change or development of an SoS, resulting from changes in its environment or in itself. In managed SoS evolution, the modification of the SoS keeps it relevant in face of an ever-changing environment; whereas in unmanaged SoS evolution, on-going modification of the SoS occurs as a result of on-going changes in (some of) its CSs.

At the conceptual level, starting from the mission, the evolution meta-requirements and the conceptual model, a set of evolution requirements produced. The latter are exploited along with the SysML profile and the building blocks at the logical level. In particular, instantiation of the profile is connected with the Structure viewpoint of the SoS. Interactions elicited among CSs take into account the service provided at the RUI interfaces, and the business value improved by evolution. The logical level results in the platform independent SoS architecture.

The role of the implementation level is to translate the generic SoS architecture into a platform-specific SoS architecture with, among others, evolution aspects. This includes RUI modification, and has a tight connection with the time viewpoint to ensure backward compatible evolution versions. Through the architecture as specialized by the enterprise it is possible to provide support to evolution though a platform-specific architecture.

Viewpoint of Dependability and Security. Dependability and security are essential properties of an SoS since they affect its availability, reliability, maintainability, safety data integrity, data privacy and confidentiality.

The conceptual level, build the set of dependability and security requirements from SoS mission, meta-requirements and the conceptual model. Dependability and security are important to ensure the proper functioning of an SoS. At the conceptual level, the input is the overall objectives and functionalities required to meet the mission of the SoS. Dependability and security requirements are not stand-alone requirements; they are connected to the other requirements, including time, multi-criticality, and others, that compose the set of requirements for the SoS.

At the logical level the Dependability and Security requirements are exploited to define the dependability and security components of the SysML profile There are two packages: "*SoS Dependability*" and "*SoS Security*". One of the key concepts in SoS dependability and security is splitting functionalities into well-defined components and interfaces such that the number of components that require explicit trust is kept to a minimum. In the context of the SysML profile, each block in the Block Definition Diagram has interaction points for *itoms* flowing in and outside the block. We first consider the functionalities required by the SoS and determine how security-critical each functionality is. We then consider what kinds of components make up the SoS and map functionalities to components. The most security-critical functionalities should be grouped together. Thus, the SoS will have a small number of highly-trusted, security-critical components. Less security-critical functionalities will be handled by less secure components. There will be different levels of dependability for each CS and different levels of security for each SoS.

The output of the logical phase is a platform-independent SoS architecture. The latter is exploited at the implementation level to create a platform-specific architecture specialized by the enterprise with their adopted technologies. For defining platform-specific trustworthy CS one could rely on the trustworthiness-enhancing design patterns described in the OPTET project [14]. This comprises a number of UML Patterns, which from the AMADEOS perspective can be seen as Dependability and Security architectural patterns.

Viewpoint of Time. Time does not only play an important role in the control of the physical environment of an SoS, where, for instance, the temporal properties of a control loop impact the efficiency and quality of control. It is also crucial for the information exchange between CSs, as in many cases timeouts and communication delays may decide whether the distinct CSs are able to serve their purposes. Correct handling of time enables the reduction of cognitive complexity required to design an SoS and facilitates the integration of new CSs into the system. On the other hand, undefined timing of communication between CSs might introduce unintended emergent effects.

Time meta-requirements along with related concepts and the SoS mission are exploited at the conceptual level to generate time dependent requirements. System components and functionalities sensitive to the progression of time need to be identified and the requirements on their temporal behavior have to be specified. This mainly comprises requirements on timeliness of interactions between CSs (e.g., the exchange of information to avoid collisions between cars has to take place before the cars collide), and the time synchronization of those CSs (e.g., requirements on the precision of synchronization and time granularity). Since there is a close relation to other viewpoints, like Security, Dynamicity or Emergence, the temporal requirements have to be aligned with the requirements regarding the other viewpoints. Furthermore, the behavior of the SoS in case that some of the temporal requirements cannot be fulfilled has to be specified.

At the logical level the temporal behaviour of the SoS is designed based on the conceptual requirements defined in the level above. The mechanism to achieve a synchronized global time base among all CSs has to be defined (e.g., internal or external synchronization of time). Such a time base allows relating timestamps of different CSs with each other, and thus enables the temporal ordering of events in the SoS. The exact temporal interaction between individual types of CSs is modelled and included in the SysML RUMI specification. A precise temporal specification at this level simplifies the integration of CSs that have been individually designed and implemented at the next levels.

At the implementation level the producer of a CS brings the temporal specification of interactions between CSs into a real implementation using a specific platform. This includes implementing the time synchronization mechanism defined in order to achieve a common time base. As the implementation has to comply with the temporal model of interactions, unintended side effects of temporal misbehaviour are avoided, and hence, the integration of the CS into the SoS is simplified and the instrumented SoS instance is created.

Viewpoint of Multi-criticality. Multi-criticality supports the provision of services of an SoS with different criticality, such as safety-critical and non-safety-critical. Indeed, while some part of the SoS may have strong safety-critical requirements, other parts may be not so critical.

At the conceptual level the definition of multi-criticality requirements is carried out in order to support the definition of services with different criticality levels. To this end, the meta-requirements is exploited according to the SoS mission and using the related SoS concepts.

At the logical level the requirements along with building blocks and the profile are exploited to define the platform independent SoS instance. The SoS architecture and RUMI specification is done so that, recalling the macro-level of the general architecture of an SoS [13], CSs characterized by a specific criticality level n and a macro-level m can rely on CSs characterized by a criticality level greater or equal to their one owned by the same or a lower macro-level.

As stated by requirement [MULTI-CR6] a CS shall not rely on CSs characterized by a lower criticality level than its one. Thus, it is also necessary to have designed a clear architecture profile which details the structure of the SoS, detailing the interaction among the CSs. In this way it is possible to verify the correctness of the interaction among the CSs checking for violations of the aforementioned requirements. In the case that a CS offer several services that are characterized by different criticality levels, then a precise specification of the RUMI building block can help to preserve both the FCR and ECR, making failure propagation from non-critical services to critical one impossible.

At the implementation level the SoS Logical description (platform independent) is specialized by exploiting the enterprise-specific technologies based on specific enterprise technologies and it will result in a platform-specific instrumented SoS architecture and RUMI specification.

Viewpoint of Emergence. Emergence is an intrinsic property of the SoS and it concerns with novel phenomena that manifest at the macro-level (i.e., at SoS level) which are new with respect to the non-relational phenomena of any of its proper parts (i.e., CSs) at the micro level. The rationale behind emergence is that by composing CSs, either positive or detrimental global emergent phenomena may occur. Managing such phenomena can help avoiding unsafe unexpected situations generated from safe CSs, and may help eliciting positive emerging phenomena.

Appropriate effort shall be devoted to *monitoring*, *analysing* and *predicting* detrimental emergence phenomena and to *mitigating* (executing appropriate reactions) their effect on the SoS. For non-detrimental emergence, it is desirable, but not mandatory to monitor, analyse and predict emergence phenomena. Emergence may be influenced or generated by modifications to the Structure (e.g., adding new components which introduces new functionalities, or adding new components that may change the error model, e.g., introducing new Itoms which enables new interoperability between CSs), dynamicity and evolution (making the system able to make changes to the way its CSs interacts with each other and how the system is aligned with changing business requirements). Note that emergence phenomena may *cause violations* to handling of time, dependability and security of the SoS/CS.

In the Conceptual level, starting from the SoS mission, the meta-requirements and basic SoS concepts, we identify the instantiated emergence requirements.

The Logical Level concerns with applying the profile to identify emergence and categorize it according to the strength and predictability of effects. Because of the nature of the emergence concept, in we deemed not sufficient to simply elicit an emergent behaviour. We also consider worth capturing operational aspects related to emergence by considering an SoS in action. For these reasons, in we consider two possible diagrams to represent emergence through Block Definition Diagram and Sequence Diagram. The building blocks defined in Section Y support the monitoring, analysing, planning and executing mitigating activities required by the emergence management requirements. The instantiation of the profile should be tightly connected with the Structure viewpoint of the SoS. Interactions elicited among CS should be defined according to the *Request-Response* model and take into account the service provided at the *RUI interfaces* as regulated by the SLA. For supporting early identification and mitigation of emergence, particular attention has to be devoted to the interactions through stigmergic channels. The design process will also consider application and domain specific details which will be added by the designer. Finally, validation activities will check the correct application of building blocks, their integration and the usage of SoS domain specific concepts (possibly available through an SoS profile).

The platform independent architecture resulting from the logical is instantiated, configured and to linked the architectural elements, which support the achievement of the emergence requirements through the implementation of a platform-specific instance.

Evolutionary Aspects. A SoS evolves over time as constituent systems are modified, replaced or added, or due to its relevant environment (gradually) changes. This evolution is driven by incremental, new, and changing requirements of the SoS. An architectural framework for SoS should provide a tool aimed at predicting possible evolutionary paths based on anticipated requirements and use-cases.

Scenario-Based Reasoning for SoS Architecture Design. In architectural systems engineering the use of scenarios is not uncommon. It is a cost-effective means of controlling risk and to maintain system quality throughout the processes of software design, development and maintenance [15, 16] Preparing for evolution of an SoS, a scenario-based approach can also be adopted to guarantee that the development that an architecture undergoes is sensible, i.e. it must guarantee that the quality goals of the system are still met.

By using scenarios to guide the design of an SoS architecture, the context of the envisioned SoS is incorporated into the possible design choices by the architect. Established scenarios provide a narrative, which enables communication about future requirements and capabilities between different stakeholders [16]. Scenario-based design is a user-based approach in which different use-cases of a system are defined by narratives, from which a lower-level description of the system can be extracted. However, not every SoS can be described by narratives focused around use-cases and user interactions. Moreover, a narrative provides the intended use of a system from the perspective of a single expert or end-user, whereas in the context of SoS single use-cases are more related to the constituent systems than to the SoS as a compound structure. Therefore, a more methodical approach is needed, in which multiple experts can define relevant states and variables that may describe the possible evolution of the SoS and its relevant environment.

Scenario-based reasoning (SBR) [17] provides a methodical approach to generate and explore scenarios. In the SBR approach, scenarios are built from a set of variables, and each combination of variable states makes up a single scenario. Relevant scenario variables are those that influence the design of the system, such as variables that denote for example: environmental conditions, organizational dynamics, economic conditions, technological development, and interactions with the system form a user perspective. Such variables can have dependence relations between them which are, for example: causal, functional, influential, or probabilistic. For instance, enabling a certain security feature in the system will typically have an influence on its usability.

SBR enables what-if exploration to reason about possible future conditions and consequences for the architecture of an SoS. Through the analysis of different scenarios and their dependencies, inconsistencies can be revealed that may have consequences for the eventual architectural design of the system. Through the identification (also generation) of scenarios from a model describing the context under which the SoS will be deployed and the possible future uses of the system, evolving requirements may be elicited. By thinking about how to operationalize these requirements, insights are acquired about how they map to the architectural design of the system.

Figure 3 shows a small sample model from an environmental point of view, from which possible scenarios can be extracted for analysis. It depicts causal relations

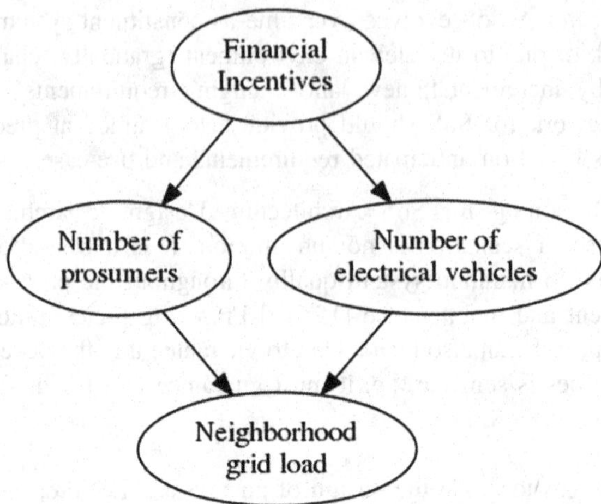

Fig. 3. A small example causal model for SBR

between the possibility of providing financial incentives for electrical vehicle use and energy production by consumers. Increased popularity of these use-cases in turn has an effect on the load placed on the local neighborhood grid.

4 The AMADEOS Building Blocks

In this section, we present the AMADEOS architectural building blocks which are exploited in the AMADEOS architectural framework.

4.1 SoS Management Infrastructure

The SoS management infrastructure in terms of a set of patterns which are applicable to enact monitoring, analysing, planning and execution strategies. The latter are developed as highly-dependable services, which we deemed essential for an SoS architecture. In order to implement the support to the above services we got inspired by the literature of Autonomic computing [18] which is a promising approach for a dependable architecture of very large information systems [3]. In particular, we propose to adopt the well-known MAPE-k cycle to implement the above services through Monitoring, Analyze, Plan and Execution components.

Our idea is to implement such patterns by means of composing CSs interacting with each other through well-defined RUI interfaces. These patterns are: (1) *Hierarchical Control*, (2) *Master/Slave*, (3) *Regional Planner*, (4) *Coordinated Control* and (5) *Information Sharing*. Patterns (1), (2) and (3) implement the so-called *Formal Hierarchy*, while patterns (4) and (5) implement the *Non-formal hierarchy*. We recall that *Formal hierarchy* and *Non-formal hierarchy* have been discussed in Chap. 3 of this book.

Formal Hierarchy. In a Formal hierarchy any CS at level n is controlled by a CS at level $n + 1$. It follows that the MAPE components are placed in the CSs forming the controlling level, i.e., level n, while controlled CSs are placed at level $n - 1$. We consider three possible instances of this pattern as follows. The *Hierarchical Control* pattern consists in having a CS implementing all the MAPE phases (see Fig. 4).

In the *Master/Slave* pattern (see Fig. 5), the controller CS implements A and P, and then delegate to additional CSs M and E (Fig. 5).

In the *Regional Planner* (see Fig. 6) the controller CS implements only the Plan phase while it delegates to a set of CSs Analysis, Monitoring and Execute phases. The CS implementing the Plan phase operates for a region of CSs for which it is responsible.

Non-formal Hierarchy. In a Non-formal hierarchy CSs at level $n - 1$ interacts with the others at the same level by creating a whole at the level n. It follows that all controlled CSs and the CSs implementing the MAPE components are all placed at the same level, i.e., level $n - 1$. Two possible implementations are as it follows.

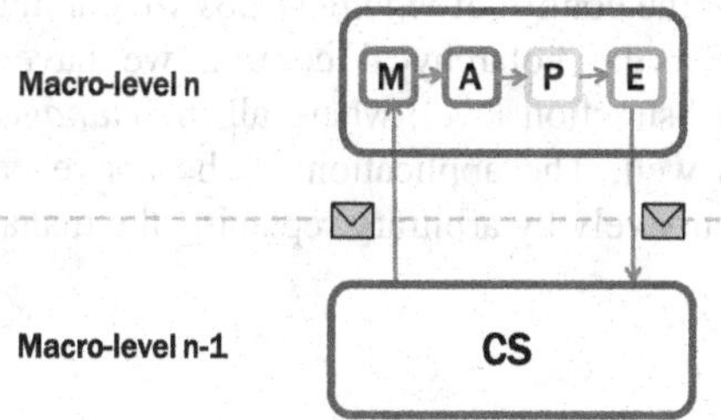

Fig. 4. Hierarchical control pattern

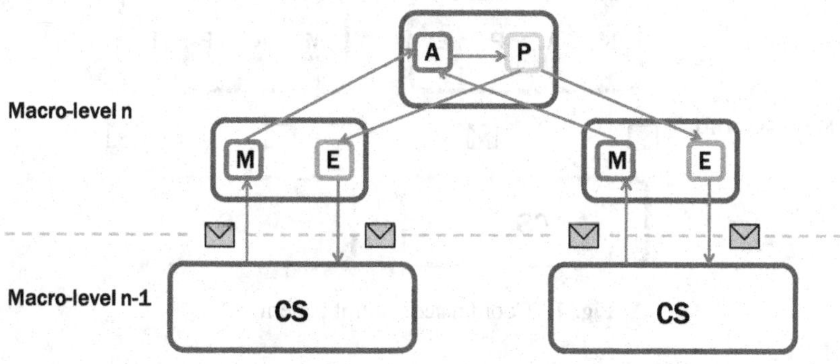

Fig. 5. Master/Slave pattern

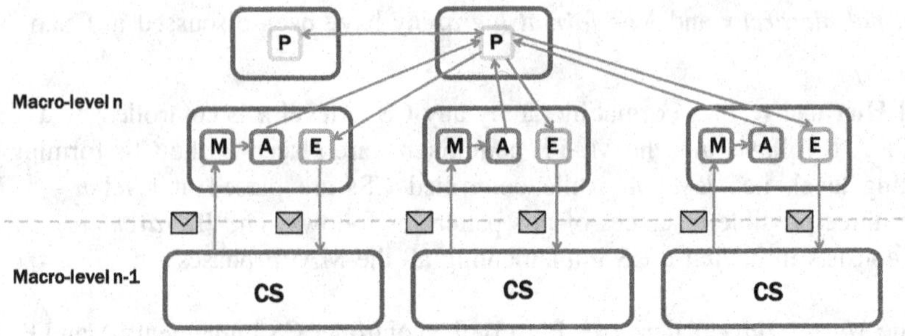

Fig. 6. Regional planner pattern

In the *Coordinated control* pattern (see Fig. 7) each of the CS at level *n* implements all the M, A, P and E phases. The latter coordinate their operation with corresponding peers of CSs at the same level (Fig. 7).

In the *Information Sharing* (see Fig. 8) is similar to the *Coordinated control* pattern but only interactions between Monitors are allowed.

Patterns composition. Each pattern presented in the earlier section exploited CSs at two possible abstraction levels. For the hierarchical control, we have at the higher level the *managing CSs* implementing the control of *managed CSs* which, in turn, have been represented as black boxes. For the holarchycal control, we have *managed* and *managing* CSs all at the same abstraction level, where all the *managed* elements are represented as black boxes, as well. The application of the above patterns may be applied compositionally and recursively by arbitrary replacing the managed CS by any other pattern.

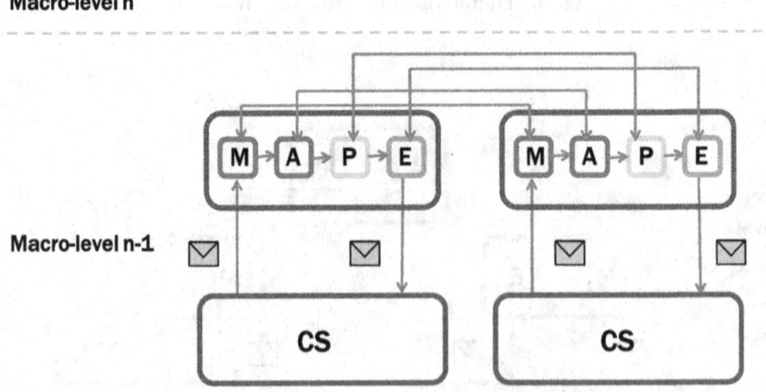

Fig. 7. Coordinated control pattern

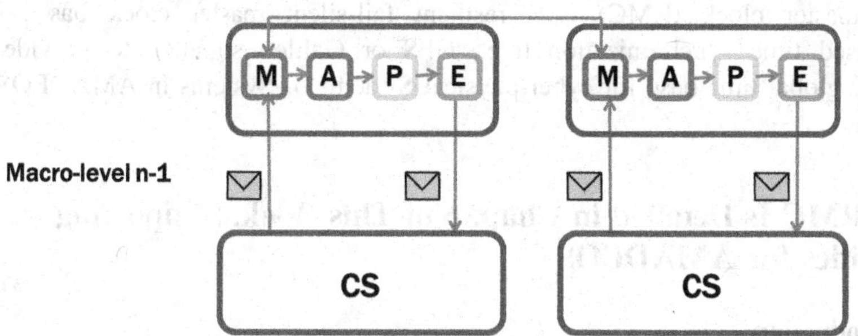

Fig. 8. Information sharing pattern

Finally, in addition to the presented patterns, a CS, being it a managing or a managed element, may interact with the physical environment by implementing the MAPE components. To this end, we introduce the atomic pattern as shown in Fig. 9. Communication Infrastructure. The communication among the MAPE building blocks is achieved by appropriate interfaces whose nature depends on the objective of the communication, either physical entities or messages. Consistently with the AMADEOS conceptual model, we adopt RUMIs to support the communication among MAPE blocks for managing SoS, since we only require the exchange of information, i.e., Itoms, and not physical entities (which would require RUPIs). Indeed, in the presented management infrastructure, our MAPE blocks do not receive physical entities but simply messages, which can be sent/received within a single CS or across CSs. Those messages have been graphical represented in the pattern as yellow envelope items. The only exception is the atomic pattern, which supports the interaction with the physical environment and consequently it requires the adoption of RUPIs to exchange physical entities. Noteworthy, we only represent RUIs to support the communication of MAPE blocks, which span different CSs while we neglect to consider MAPE interactions within a single CS.

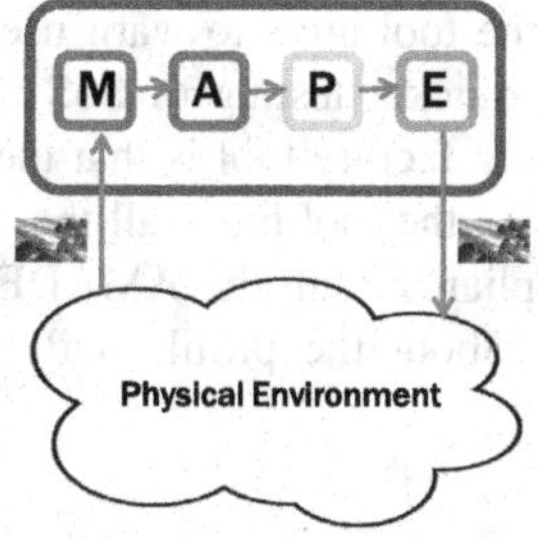

Fig. 9. Atomic pattern

4.2 Resilient Master Clock

Resilient master clock (RMC) is a resilient fail-silent master clock based on satellite-based time synchronization (e.g., GPS or Galileo signals), to provide a dependable global time base for cyber-physical Systems-of-Systems in AMADEOS.

5 The RMC Is Detailed in Chap. 6 of This Book. Supporting Facilities for AMADEOS

5.1 Introduction

The *supporting facility tool*[1] is used to model, validate, query, and simulate an AMADEOS based SoS using the Blockly tool[2]. Blockly is an open source library for building visual programming editor or a visual DSL (domain specific language). Blockly has been adopted to ease the design of SoS by means of simpler and intuitive user interface; thus requiring minimal technology expertise and support for the SoS designer. Its main features are: (i) Fast, and only a modern web browser is required; (ii) Intuitive and simpler user interface; (iii) Easily extendable with custom blocks; (iv) Ability to check constraints at design time (user defined and pre-defined constraints) and warn user when the user makes mistakes; and (iv) Support code and XML generation.

The supporting facility tool is a generic SoS designer in accordance with the AMADEOS conceptual model and for this the Blockly tool has been customized to be used for SoS modelling. The flow of model-driven engineering using the supporting facility tool is depicted in the Fig. 10. The SysML meta-model is first transformed to Blockly blocks. These blocks could be used in the supporting facility tool to create an SoS model.

The main motivation of *supporting facility tool* is: the current SoS design tools are complex and non-intuitive for general SoS designers; also, many of the existing tools expect designers to be well-versed with object-oriented concepts. The goal of supporting facility tool is to simplify and provide means to rapid modelling of SoS using the SysML profile (meta-model). In traditional modelling environment, large models have been known to be difficult to design and maintain; and often leading to spaghetti diagrams. The tool aims to reduce the complexity by using collapsed views instead of lines to connect blocks. Also, the tool aims to warn user of common errors/mistakes during modelling and helps in quicker testing of SoS through simulation. The main advantage of using the supporting facility tool is that the SoS designer need not have deep knowledge of SysML/UML; the tool hides all the object-oriented concepts from the user and provides full compliance with the AMADEOS profile. The only prerequisite is high-level knowledge about the profile and knowledge of the supporting facility tool usage.

---

[1] http://blockly4sos.resiltech.com.

[2] https://developers.google.com/blockly/.

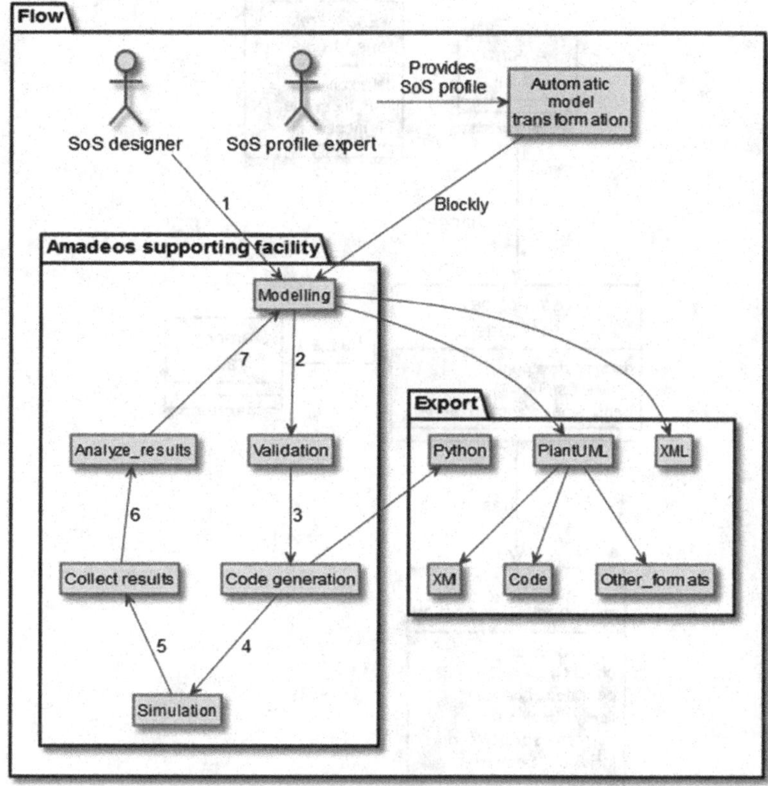

Fig. 10. Flow of MDE using the supporting facility tool

The supporting facility simplifies the task of SoS modelling by reducing the pre-requisites to start modelling. Once the supporting facility is installed on a web server, it can be accessed from any machine using a modern web-browser. It can also be used locally without the need of a web server. It provides rapid modelling, validating, code-generation, and simulation facilities to the user. The supporting facility can generate three outputs: (i) the model in XML, (ii) Python code-generated for the simulation, and (iii) PlantUML version of the model.

PlantUML is a simple text based UML format which can be readily integrated with many tools[3]. The exported model in PlantUML may be used for further refinement or formal analysis. For example: the PlantUML model can be *viewed* in Eclipse using plug-ins[4]. Though, full interpretability between tools is an ongoing research topic and is under investigation.

[3] http://plantuml.com/running.html.

[4] http://plantuml.com/eclipse.html.

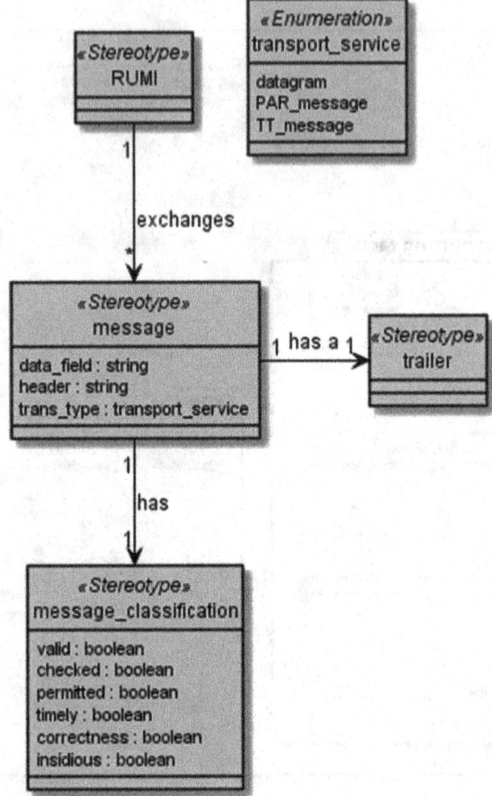

Fig. 11. An example subset of SysML meta-model to be transformed to Blockly

Python is a general purpose portable language, and the Python code generated by the tool can be further refined and also be used to connect to other simulators or external systems for interaction while running simulation.

As the supporting facility tool is based on the SysML profile (the meta-model) derived from the AMADEOS conceptual model, the SysML (in XML) is transformed into Blockly by using PlantUML as an intermediary language. PlantUML is chosen as an intermediate format as it is a simple text format which makes debugging during the model transformation easier. Below is an example of model transformation from SysML in Papyrus/Eclipse to Blockly (Figs. 11 and 12).

5.2 Modelling SoS

When the tool is launched, it creates a default SoS block called "*example_block*" as an example. All the blocks required to build an SoS can be found in the toolbox on left hand side. These blocks are imported from the AMADEOS SysML profile provided by a profile expert. Each block in the tool contains information taken from the AMADEOS

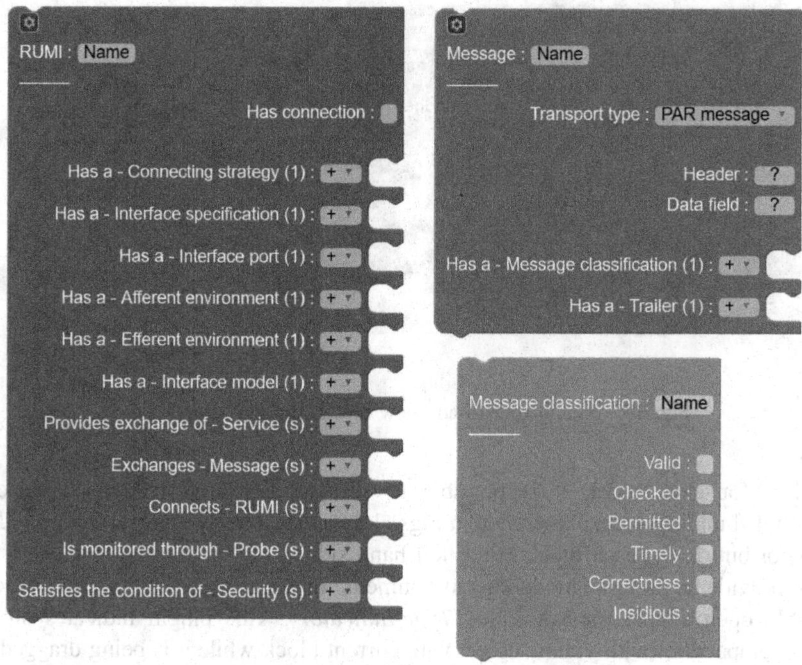

Fig. 12. SysML (Fig. 11) imported to Blockly

conceptual model to guide the SoS designer. For example, help for CS block can be found by right clicking a block and selecting *Help*. Also, each imported block in Blockly is associated with a viewpoint/building-block, for example all blocks associated with Communication viewpoint is present in the Communication category in the toolbox.

Traditionally, Blockly requires users to drag and drop blocks from flyout/toolbox to create new blocks. To improve usability and correctness, a Blockly API*: Blockly. FieldDropdown()* is used to show the list of blocks compatible to be connected for a given block; this lets the user create blocks in an easier way. Figure 13 shows an example, where to add a *Technique* block, the tool shows that the following new compatible blocks can be added: "*Fault forecast*", "*Fault prevention*", "*Fault removal*", and "*Fault tolerance*". In the profile, Technique is an abstract block and the above four blocks inherit the Technique block.

A block once created can have three views, (i) collapsed view, (ii) partially-collapsed view, and (iii) uncollapsed view as shown in Fig. 14. Collapsed view allows the user to reduce the number of blocks screen on the screen and to focus on the current editing block. Partially collapsed block only shows the non-empty attributes of a block hence the designer may choose to view only the attributes defined. Full view/uncollapsed view is used to see all the attributes of a block. A user can cycle between the three views by double-clicking the block.

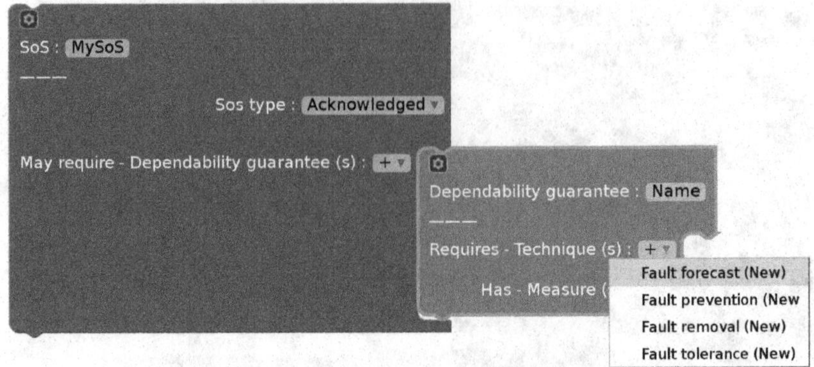

Fig. 13. Aiding user to add new blocks through dropdown

Also, for each block it is possible to see the attributes related to selected viewpoints/building-blocks as shown in Figs. 15 and 16. This is achieved by providing a mutator button for each block at top left hand side.

To provide an intuitive modeling environment, the supporting facility uses a readily available open source plug-in called *Type-Indicator*[5]. This plugin indicates all the blocks compatable (with yellow color) with current block while it is being dragged, as shown in Fig. 17 (the block *cs4 is* currently being dragged*)*.

Requirements Management. Requirements management is an important aspect of an SoS design, where traceability of requirements must be viewed/monitored. Requirements may be divided based on the viewpoints and building-blocks: Architecture, Communication, Dependability, etc. Each block maintains the list of requirements it meets and each requirement block maintains the list of blocks which satisfy it; thus offering full traceability (Figs. 18, 19, and 20). Blockly also supports adding comments to blocks to make the design clearer.

Constraints in the Model. Each block exports a list of variables in *JavaScript* which can be used to define constraints. These variables are defined in the format: *block.<relation_name>_<block_type>* (For e.g. a CS block exports block.provides_service). Also, each block exports shortcut variables in the form *block.m_<block_type>* (e.g. for a CS block, block.m_service). Instead of *"block"* keyword, a shortcut variable "*b*" may also be used.

For multiple inputs, a dictionary variable in the form *"d_<variable_name>"* is also exported. This variable is used to access variables by using block name as a key (e.g. for an SoS, Using the variable *block.d_cs['cs1']* the CS in the SoS having name *cs1* can be referred. Constraints make a model precise, the constraints provided by the tool uses JavaScript's "*eval*" function to evaluate the constraints and change the color of block to *black* in colour if the constraint is not satisfied. The constraints are evaluated at

[5] https://github.com/HendrikD/blockly-plugins/tree/master/type-indicator.

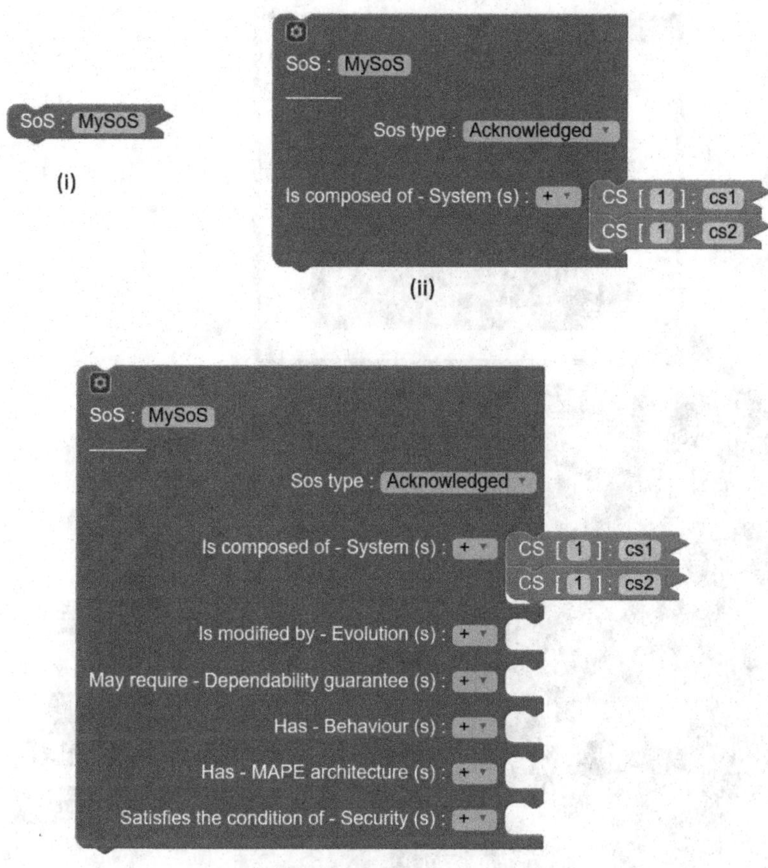

Fig. 14. Three ways to view a block (i. Collapsed, ii. Partially-collapsed, and iii. Un-collapsed)

each *onchange* event of block. Constraints rely on the variables exported by a block. Figure 21 shows an example use of constraints.

Constraints may also be used to detect causal loops which may lead to emergence scenario in SoS (Fig. 22).

Model Querying. On large models it is difficult to visualize the entire SoS, and then the need for custom viewpoints arises. Blockly does not use lines to show relationship between blocks and uses collapsed views to hide the complexity of an SoS model. Model querying can be used search for blocks which satisfy a given condition (using a query). It may also be used to visualize a model in traditional view (i.e. showing blocks and its relationship with other blocks using lines). To query a given model, a user can right click on workspace and choose "*show query diagram*. In the query diagram, user may write a filter function for querying the model. For example, *return true;* indicates that no filtering is required (i.e.: show all blocks for the model depicted in Fig. 23); which results in the graph as shown in Fig. 24. Using the filter "*return b.*

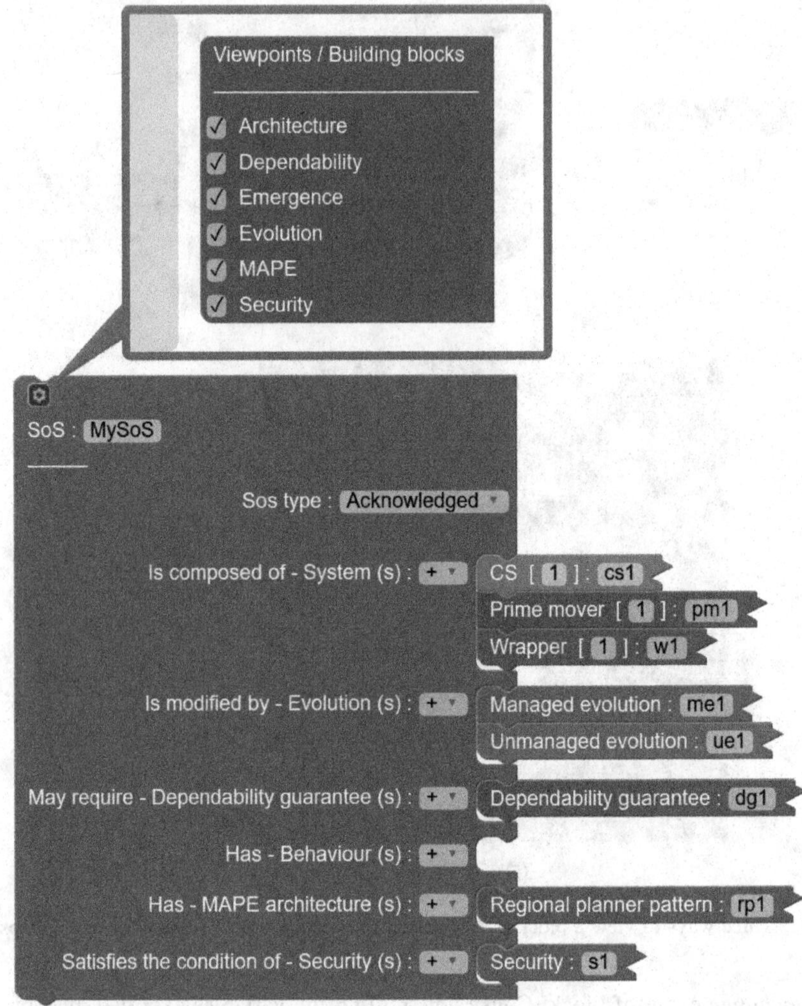

Fig. 15. Viewpoints/building-blocks of a block can be enabled or disabled

of_type == 'RUMI';" which indicates to highlight all blocks of type *"RUMI"*, this query returns the graph depicted in Fig. 25 (note that *b* is a shortcut for variable *block*). Model querying helps in visualizing custom viewpoints of SoS and can be helpful in identifying issues in the SoS design.

Adding a Link to a Block. One way to design a SoS is by using links to existing blocks. Creating links can help reuse an existing block; however, this is different from copy-pasting a block in blockly. Links are reference to the linked blocks. For example: CSs can be created on workspace and only links may be added to the SoS block, as shown in Fig. 26.

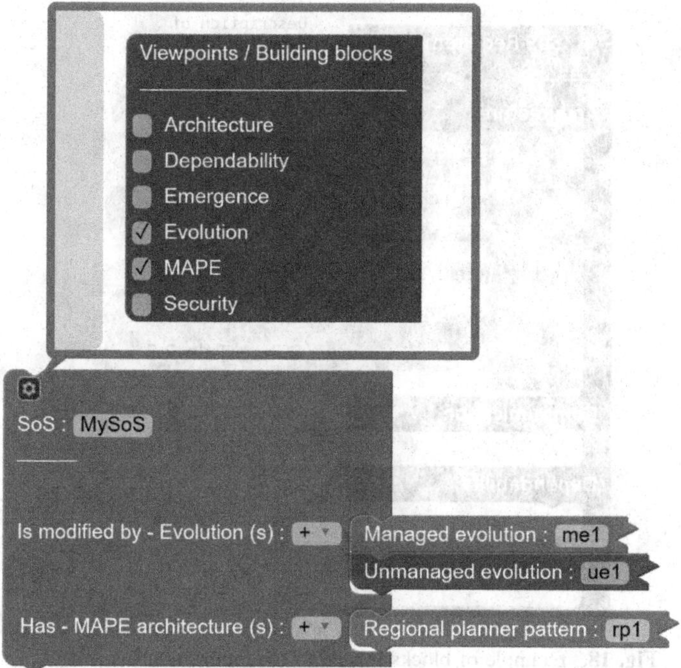

Fig. 16. Filtered view of SoS

Fig. 17. Use of Type-Indicator Plug-in (compatible connections for cs4 are indicated by yellow colour) (Color figure online)

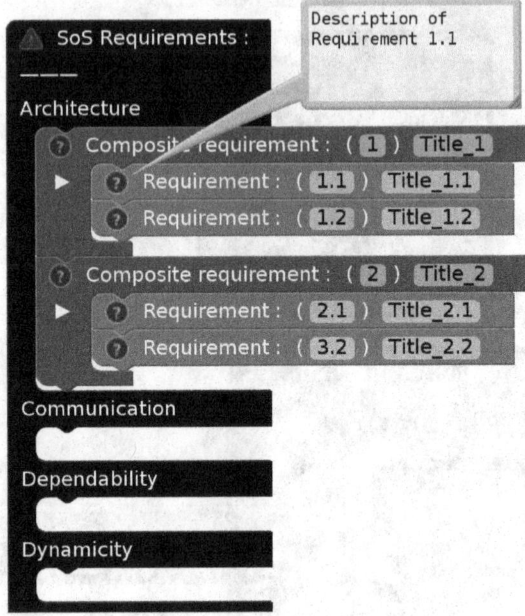

Fig. 18. Example of blocks related to requirements management

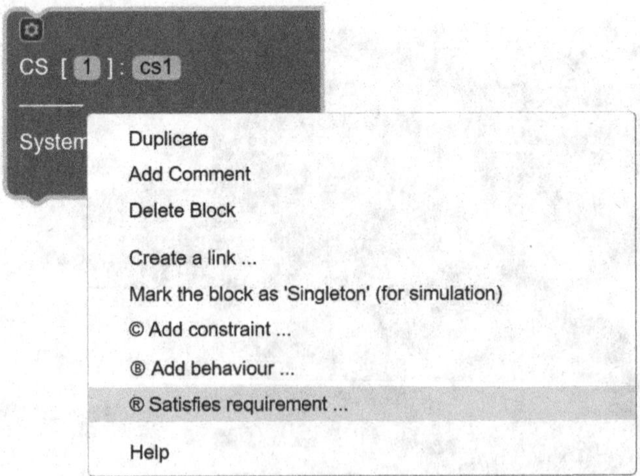

Fig. 19. Each block can satisfy a requirement (by providing the requirement ID it satisfies)

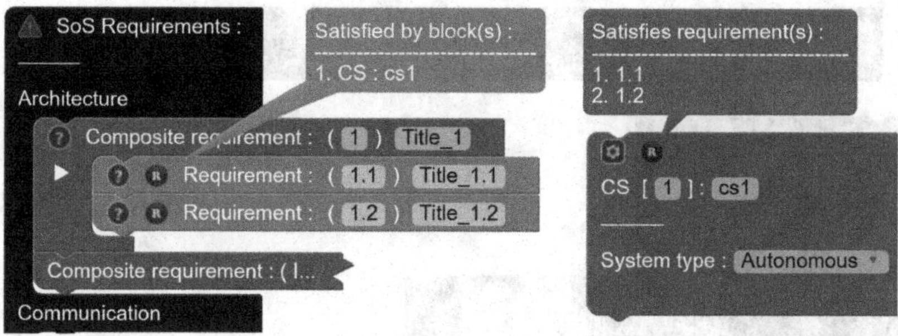

Fig. 20. Traceability of requirements

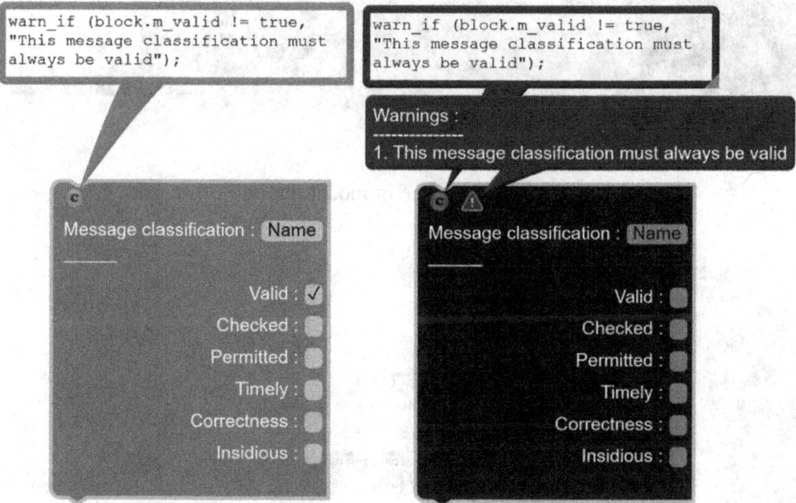

Fig. 21. An example of a constraint where the member variable m_valid is checked

Grouping for Modular SoS Design. The supporting facility allows grouping of compatible blocks together to modularize the design. For example, all CSs can be grouped together as shown in Fig. 27. The group block helps in organizing the model into meaningful groups. Also, when a block of a group blocks is refered, the group name is indicated to distinguish it from other blocks which may have similar names.

5.3 Simulation Environment for SoS

Behaviour. Once a static model is defined, behaviours may be added to any block. To add a behavior, the user can right click on the interested block, and choose "Add behavior". The behavior represents the code to be executed during simulation, and can be written in Python programming language (as shown in Fig. 28). The function names

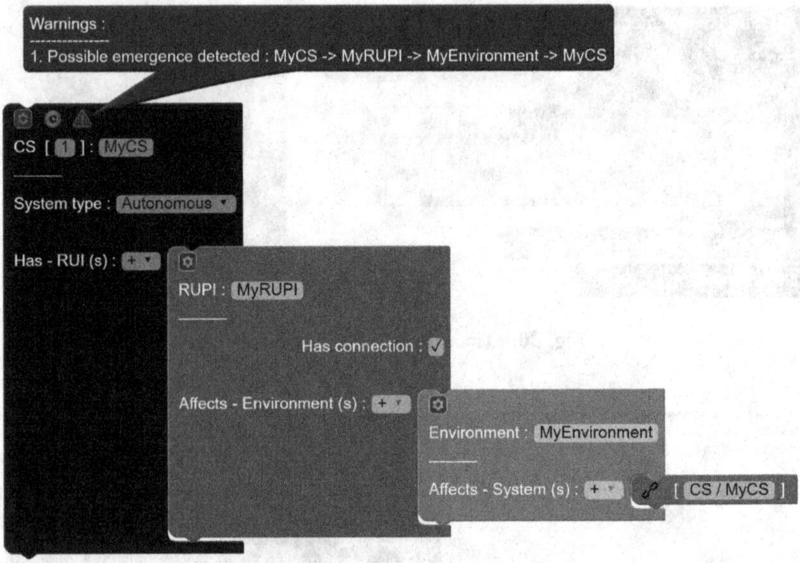

Fig. 22. Detecting emergence in model through constraints

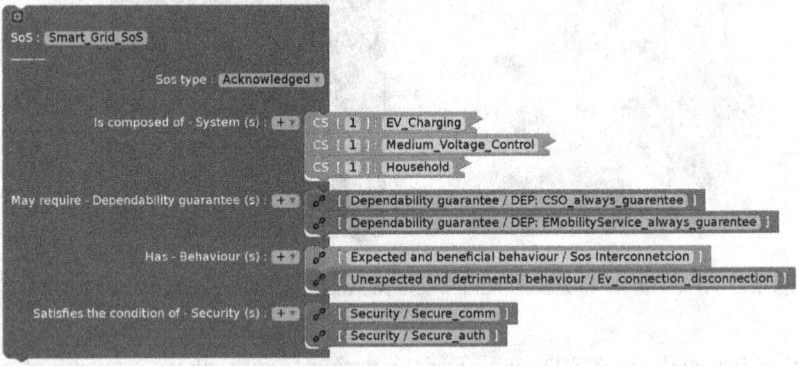

Fig. 23. Model querying large models (for query "return true;" i.e. show all blocks)

init, *start*, and *run* can be defined and are executed during initialization, start of the block, and during the course of simulation respectively.

The *run* function for a service block has a special meaning and is exposed as a TCP/IP server. All the behavior code written for all blocks are integrated in to a single file for code generation.

XML and Code Generation. After the model is loaded, it can be exported to XML and code for simulation by clicking on the appropriate buttons on the top right hand side of

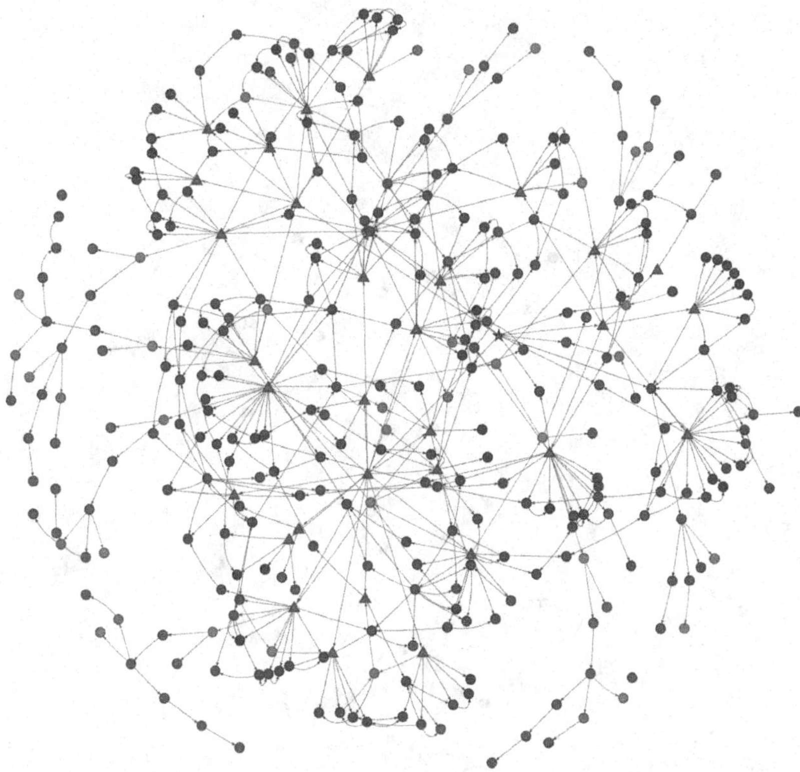

Fig. 24. Result of "return true;" query

the tool. Unique object names are generated for all blocks in a format: *<block-type>_<block-name>_<block-id>*.

Simulator Components. The simulator is a set of Python programs meant for executing the desired scenarios created by designer (the scenarios may also be represented using sequence-diagrams). The simulator consists of the following main components: Object initializer, Registry, Sequence diagram, GUI, Runtime sequence diagram, **log** generator, and Clock.

Object Initializer. The simulation initializes each object/block defined in the model using the block's constructor. Single inputs are considered as strings/integers/object; whereas multiple inputs are considered as array. If a value for single input is not provided, its value is considered as *None* in Python; whereas for multiple inputs it is considered as an empty array *[]*.

Certain blocks such as CS/Wrapper/Roleplayer/CPS can have a member called *"cardinality"* (Fig. 29). It indicates number of objects to be simulated. This is implemented by using *copy.deepcopy ()* function of Python on the original object. Each instance is assigned a *_instance_id* (1 to N); where *N* is the cardinality specified in the

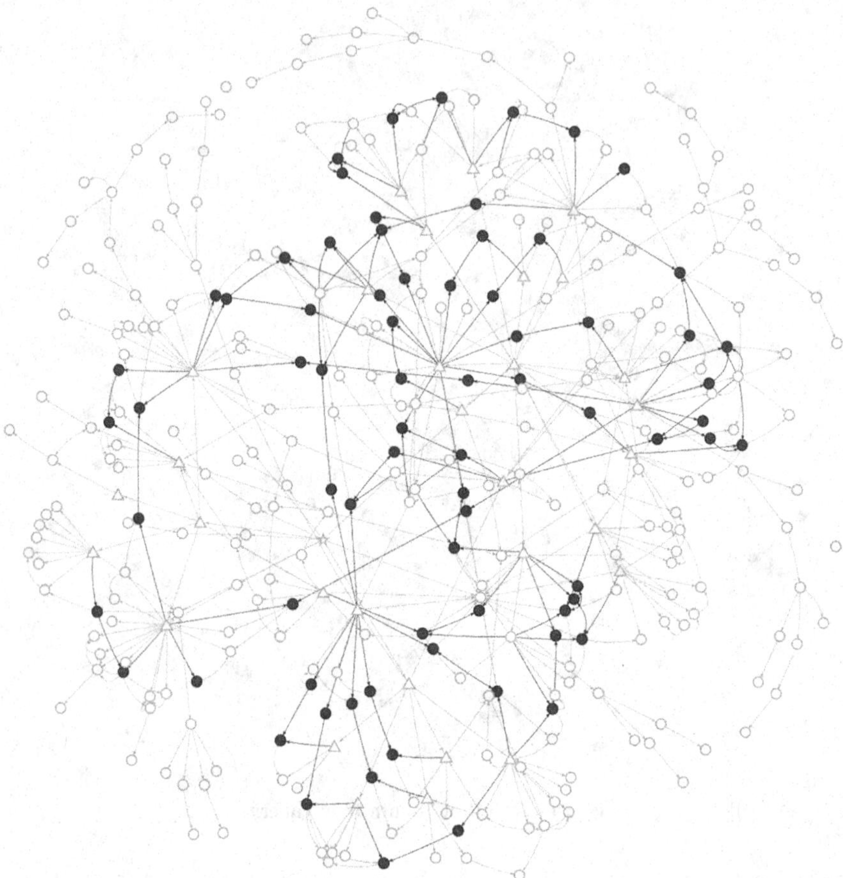

Fig. 25. Result of "return b.of_type == 'RUMI';" query (select all RUMIs)

model. Example: the below model creates an SoS called MySoS and has 200 CSs having name cs1. Each of the CSs will have *_instance_id* attribute from 1 to 200.

Registry. Registry is one of the main components of the simulator. It is a service that maintains the list of services offered by various CSs registered in a SoS. It is used by the CSs to search for a particular service. In the simulator, the registry is implemented as a TCP/IP server, where CSs can add/remove/update their own service information. Having a known common registry allows the possibility to run the simulation across several computer systems connected together.

Sequence Diagram. Blockly blocks related to sequence diagrams helps to create non-ambiguous sequence diagrams, which can be readily converted to code. Simulator follows the exact sequence as defined in the sequence-diagram created by the user. Thus, the code generated from the sequence diagram (Fig. 30) is executed right after the simulator has been started and initialized. A sequence diagram is added to model to simulate a scenario (Fig. 30); the sequence diagram designed in supporting facility tool

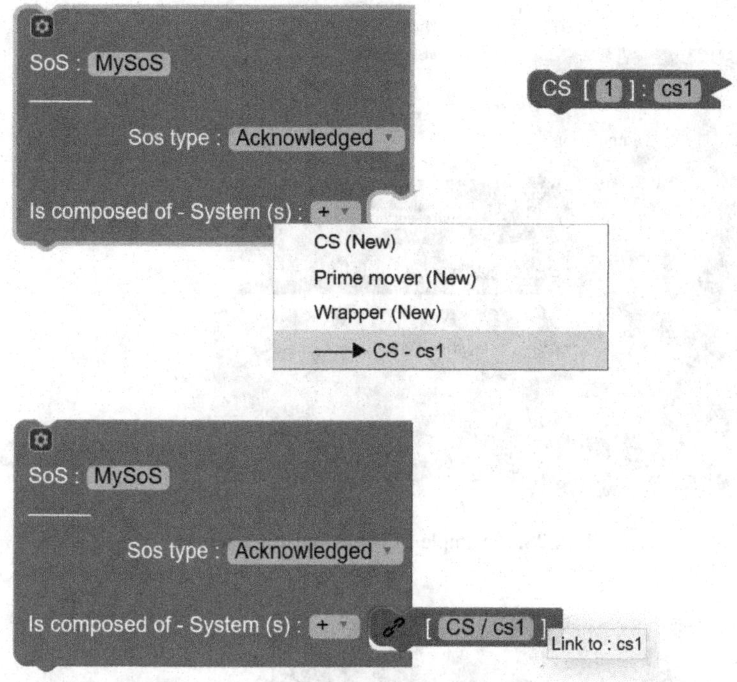

Fig. 26. Reusing an existing block (cs1) using links.

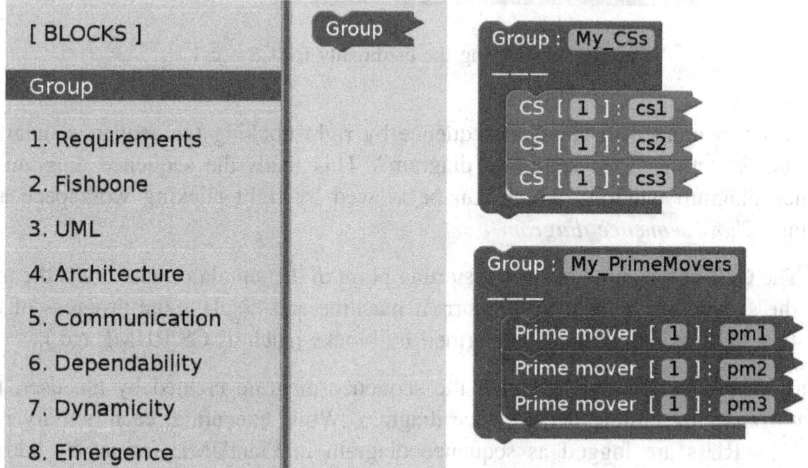

Fig. 27. Similar blocks can be grouped together

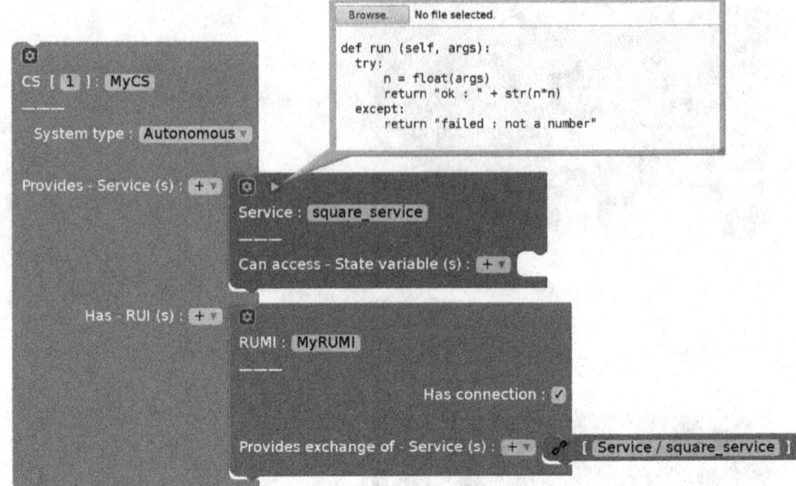

Fig. 28. Example of behaviour for a service

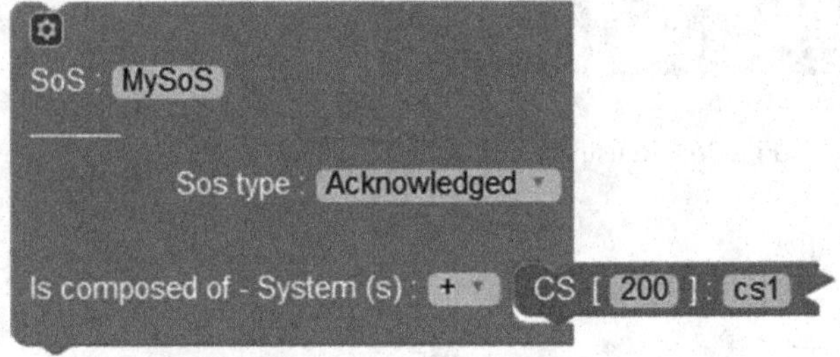

Fig. 29. Specifying the cardinality for CS – cs1

can also be visualized in classical sequence by right clicking the sequence diagram block and selecting "load sequence diagram". This loads the sequence diagram in sequence diagram window, which can be viewed by right clicking workspace and selecting "*Show sequence diagram*".

GUI. The GUI of the simulator is the starting point of the simulator, and it lets the user select the systems to be run on the current machine and displays the progress of the simulation by logging activities performed by blocks (such as CS/RUMI, etc.).

Runtime Sequence Diagrams. Given the sequence diagram created by the user, the simulator starts executing the sequence diagram. While executing, each activity performed by RUIs are logged as sequence diagram in PlantUML format by adding timestamp to each activity. This creates a runtime-sequence diagram *(in result.seq file)*, which shows what actions have occurred with its timestamp. The runtime-sequence diagram also shows the delay between each action.

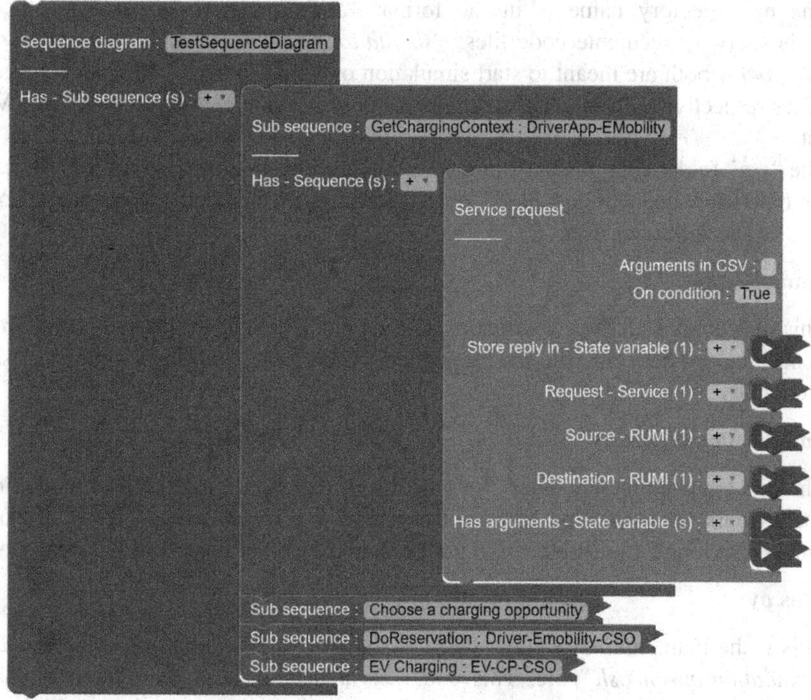

Fig. 30. Sequence diagram in supporting facility tool using Blockly

Log Generator. The logs generated by the simulator can be saved in a file and can be used to compute the metrics of interest. These metrics may indicate the quality of SoS by measuring performance/delays/failures etc.

Clock. The simulator uses the system clock of the machine on which the simulation is running. However, it is possible to setup an experimental setup in which, each CS runs on different machine using different clocks. These clocks could be synchronized with a master clock e.g. the RMC developed in task D 4.4 [19]. Faulty scenarios (regarding time synchronization) are also possible to generate by perturbating local clocks of each machine, or by removing master clock from the network.

Simulator Code Organization. The code generation of the supporting facility tool generates a ***".zip"*** file in the format *"<model-name>.zip"*, which contains the complete code for the simulation. The simulator code is created for each model based on the specified sequence diagram. The generated code when extracted is organized as shown below (the model name is *"sos-model"*):

The top directory name is in the format *"SoS-Simulation-<Date-and-Time>"*, which hosts two executable code files: *"simulation-on-unix.sh"* and *"simulation-on-windows.bat";* both are meant to start simulation on UNIX-based and Windows-based machines respectively. The *model-<Date-and-Time>.xml* file consist of model in XML format.

The "*src*" folder contains the simulator code, the constructor code for each blocks, and initialization code of the block objects created in the model. The entire code consists of the following files:

(1) **amadeos.py**

This file contains all the constructors for each block defined in the SoS profile from the conceptual model. This file may be edited to add/refine additional generic classic functionalities.

(2) **model_behaviour.py**

This file contains the behaviors for each block defined during the SoS modelling. The behaviours are associated with each instance of a block and not for each class. Thus the behaviour for one object will not be shared by other objects of the same class.

(3) **sos.py**

This is the main simulator code which is started by *"simulation-on-windows.bat"* or *"simulation-on-unix.sh"* file. This code sets a random seed for random number generation, creates the registry, sets global data for simulation, and starts the user interface for simulation.

Also, this file contains code that starts the simulation by starting all systems as a thread, runs the code related to the sequence diagram, and waits for all threads to join.

(4) **sos_gui.py**

This file contains the GUI code for selecting the systems to be started on the running machine. Also, this file contains code for showing the log of activities performed by each CS.

Running the Simulation. After the code generation, a user can start the simulation by launching the file "*simulation-on-unix.sh*" or "*simulation-on-windows.bat*" (Fig. 31). When the simulation starts, a GUI is shown that allows the user to select the list of systems to be started, and the registry IP address and port. An example GUI is shown in Fig. 32. After selecting the list of systems to be started, the user can start simulation by clicking the "*Start simulation*" button.

Simulation Over a Network of Computers. The SoS simulation may also be performed over a network of computers. This is achieved by maintaining a common registry machine; thus forming a distributed system running various AMADEOS based systems in each of the computer systems communicating through TCP/IP.

This also allows the possibility of the simulator to interact with real legacy-systems. Each machine can run a set of systems (CSs/Wrappers/Primemovers). When run, a

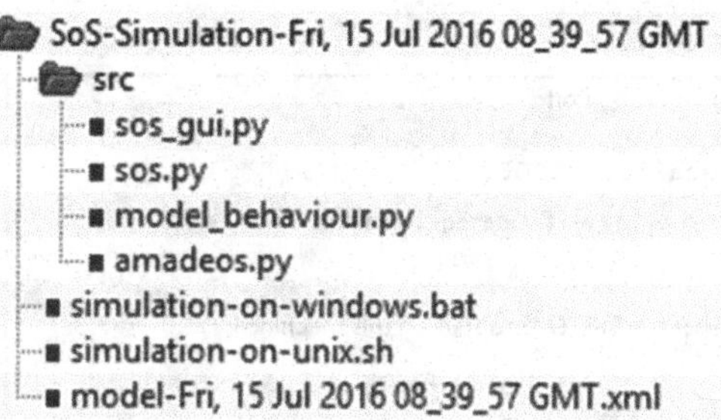

Fig. 31. Simulator code directory structure

separate thread is created for each selected system; and each system initializes itself and starts its RUIs in separate threads.

Example Run of an SoS Simulation. This section describes and the steps for running the simulation using an example simulation of a SoS model designed in the supporting facility tool. An example SoS may be launched from the dropdown found on the top left hand side of the tool.

5.4 Prerequisite to Run Simulation

As simulator code is written in Python 2.7, the pre-requisite to run simulation is an installation of Python version 2.7[6]. On Windows it is preferred to install Python at c:\Python27, which is the default option provided by the installer.

5.5 Starting the Simulator

As mentioned earlier, the simulator code will be generated as a file in the form *"<sos-name>"*.zip, containing a *src* folder and two files: *"simulation-on-windows.bat"* and *"simulation-on-unix.sh"*.

The simulator starts when the user runs the script *"simulation-on-windows.bat" or "simulation-on-unix.sh"* depending on the operating system.

For security reasons, on some versions of Windows it may be required to right click on the *"simulation-on-windows.bat"*, click to properties, and check "Unblock" this file.

When the simulator starts, it shows a GUI (Fig. 32), where the user can specify the list of systems to be started on the current machine. After system selection, the user may click "Start simulator" to start the simulation. The user can see the log of activities appearing during the simulation on the GUI (Fig. 32).

After the simulation run, the user may close the GUI. The simulator generates the result of the current simulation, i.e., message passing between RUMIs and interactions between RUPIs - as a run-time sequence diagram in a file called *"result.seq"*.

[6] https://www.python.org/download/releases/2.7/.

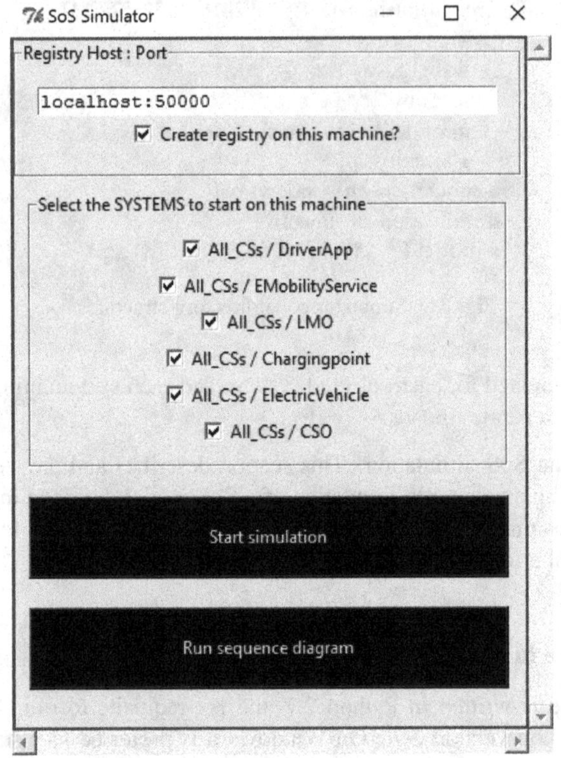

Fig. 32. Start-up GUI of simulator

The result is saved in the same directory where the simulation was run. This file can be viewed by a PlantUML viewer or a sequence diagram-viewer available in the supporting facility tool by right clicking on workspace and selecting "*Show sequence diagram*". The sequence diagram frame can be extended to fit the page, and the user can use the "*Browse button*" in sequence diagram frame to load the *"result.seq"* file.

6 Conclusion

This chapter has introduced the architectural framework and supporting facility tools for AMADEOS based SoS.

This chapter has showcased the features of the supporting facility tool, focusing on simplicity and intuitiveness in modeling and simulating an SoS. The supporting facility tool also demonstrates the possibility of: design, validation, querying, simulation of system of systems. Case studies using the supporting facility tool will be presented in Chap. 8.

References

1. Schonenborg, R., Bieler, T.M.A., Fijneman, M.: System of systems architecture in ESA's concurrent design facility. In: SECESA - System and Concurrent Engineering for Space Applications, Lausanne (2010)
2. Muller, G., Van de Laar, P.: Researching Reference Architectures. Springer, Netherlands (2009)
3. Murer, S., Bonati, B., Furrer, F.J.: Managed Evolution: A strategy for very large information systems. Springer, Heidelberg (2011)
4. Atego: Artisan Studio, March 2012. http://www.atego.com/products/artisan-studio/
5. DelCOMPASS: Guidelines for Architectural Modelling of SoS. Technical Note Number: D21.5a Version: 1.0, September 2014. http://www.compass-research.eu/Project/Deliverables/D21.5a.pdf
6. delDANSE: DANSE Methodology V2 - D_4.3. https://www.danse-ip.eu/home/images/deliverables/danse_d4.3_methodology_v2.pdf
7. Gezgin, T., Etzien, C., Henkler, S., Rettberg, A.: Towards a rigorous modeling formalism for systems of systems. In: IEEE 15th International Symposium on Object/Component/Service-Oriented Real-Time Distributed Computing Workshops (2012)
8. Rao, M., Ramakrishnan, S., Dagli, C.: Modeling and simulation of net centric system of systems using systems modeling language and colored petri-nets: a demonstration using the global earth observation system of systems. Syst. Eng. **11**(3), 203–220 (2008)
9. Huynh, T.V., Osmundson, J.S.: A systems engineering methodology for analyzing systems of systems using the System Modelling Language (SysML), Department of Systems Engineering, Naval Postgraduate School, Monterey (2006)
10. Lane, J.A., Bohn, T.B.: Using SysML modeling to understand and evolve systems of systems. Syst. Eng. **16**(1), 87–98 (2013)
11. Bohn, T., Nolan, B., Brown, B., Balmelli, L., Wahli, U.: Model driven systems development with rational products, IBM Redbooks (2008). http://www.redbooks.ibm.com/abstracts/SG247368.html?Open
12. Project AMADEOS: Deliverable D2.2 "AMADEOS conceptual model" (2015)
13. Project AMADEOS: Deliverable D1.1 "SoSs, commonalities and Requirements" (2014)
14. OPTET: OPerational Trustworthiness Enabling Technologies. FP7-ICT-2011.1.4 - Trustworthy ICT project
15. Kazman, R., Carrière, S., Woods, S.: Toward a discipline of scenario-based architectural engineering. Ann. Softw. Eng. **9**(1–4), 5–33 (2000)
16. Rosson, M.B., Carroll, J.M.: Scenario-based design. In: Jacko, J., Sears, A. (eds.) The Human-Computer Interaction Handbook: Fundamentals, Evolving Technologies and Emerging Applications. Lawrence Erlbaum Associates, pp. 1032–1050 (2002)
17. Conrado, C., de Oude, P.: Scenario-based reasoning and probabilistic models for decision support. In: 2014 17th International Conference on Information Fusion (FUSION) (2014)
18. Kephart, J., Chess, D.: The vision of autonomic computing. IEEE Comput. **1**(36), 41–50 (2003)
19. Project AMADEOS: Deliverable 4.4 - Design of Resilient Master Clock (2016)

Time and Resilient Master Clocks in Cyber-Physical Systems

Andrea Ceccarelli[1(✉)], Francesco Brancati[2], Bernhard Frömel[3], and Oliver Höftberger[3]

[1] Department of Mathematics and Informatics, University of Florence, Florence, Italy
andrea.ceccarelli@unifi.it
[2] Resiltech SRL, Pisa, Italy
francesco.brancati@resiltech.com
[3] Institute of Computer Engineering, Vienna University of Technology, Vienna, Austria
{froemel,oliver}@vmars.tuwien.ac.at

1 Introduction: Challenges for Time-Aware Cyber-Physical Systems-of-Systems

1.1 On the Role of Time in Cyber-Physical Systems-of-Systems

Since many years, it has been acknowledged that the role of time is fundamental to the design of distributed algorithms [21]. This is exacerbated in cyber-physical distributed systems, and consequently in Systems-of-Systems, where it is sometimes impossible to say which one of two observed environmental events occurred first.

Computers, and consequently constituent systems (CSs), use at least a basic mechanism for local time keeping, in the form of incremental timers. In fact, each autonomous CS has its own oscillator that swings freely and is uncoordinated with respect to the oscillation of the oscillator in any other autonomous CS. These oscillators are often adequate to make local duration measurements, generate alarms by time-out, etc. However, there are reasons for resorting to clocks that give you an absolute notion of time. Clocks are the only way to achieve tightly synchronized actions of an ensemble of nodes in a System-of-Systems [23].

For example, *synchronized clocks* make it possible to measure the duration of some action that starts in one node and ends in another node [22]. This is a common requirement in many applications, where the duration between events that occur in the environment of the different CSs of an SoS must be determined. It is not possible to measure the duration between events that occur in the physical environment of different CSs if no global notion of time of adequate precision is shared by all CSs of the SoS.

If a *global timestamp* is assigned to every significant event, then the duration between any two significant events occurring at any place within the whole SoS can be

This work has been partially supported by the FP7-610535-AMADEOS project.

A. Bondavalli et al. (Eds.): Cyber-Physical Systems of Systems, LNCS 10099, pp. 165–185, 2016.
DOI: 10.1007/978-3-319-47590-5_6

calculated easily. For example, we can consider the temporal validity of real-time data. An observation of a dynamic entity, e.g., the state of traffic light, e.g., green, can only be used for control purposes within a validity interval that depends on the dynamics of the entity (the traffic light). If the observation of the environment is performed by a CS that is different from the CS that uses the observation, then, based on the timestamp of the observation, the user can determine if the given observation is still valid to use at a particular later instant [1, 2]. Given that such a global SoS time is available, this global time can be used to radically simplify the solution of many other temporal coordination problems in an SoS [1, 2].

As an SoS is sensitive to the progression of time, such global notion of time is required in order to [2, 6], amongst other: (i) enable the interpretation of timestamps in the different CSs; (ii) limit the validity of real-time control data; (iii) synchronize input and output actions across nodes, with specific reference to *stigmergic* and *message-based* information exchange; (iv) specify the *temporal properties* of interfaces; (v) perform prompt error detection; (vi) strengthen security protocols; (vii) allocate resources *conflict-free* (e.g., in time-triggered communication, scheduling).

A *dependable global (physical) time* is thus needed to establish the backbone of the temporal infrastructure of an SoS. Every CS in the SoS that is subject to physical time requirements should be able to measure time with an appropriate precision, and achieve a quality of time synchronization which is deemed sufficient [24]. Such a dependable global (physical) time is a fundamental requirement for time-aware SoS, although we remark that it is well-known that it is impossible to precisely synchronize the clocks in a distributed computer system. A measurement error in the timestamps of events is unavoidable. This measurement error can lead to inconsistencies between the actual and recorded temporal order of events [1, 25].

1.2 Towards a Dependable Global Time-Base

In an SoS, *external clock synchronization* is the preferred alternative to establish a global time, since the scope of an SoS is often ill defined and it is not possible to identify *a priori* all CSs that must be involved in the (internal) clock synchronization. A CS that does not share the global time established by a subset of the CSs cannot interpret the timestamps that are produced by this subset. The preferred means of clock synchronization in an SoS is the external synchronization of the local clocks of the CSs with the standardized time signal distributed worldwide by satellite navigation systems, such as GPS, Galileo or GLONASS. Standalone satellite navigation systems are based on receivers processing GNSS (Global Navigation Satellite Systems) satellite signals. GNSS currently have two core constellations: Global Positioning System (GPS) of the United States and the Global Navigation Satellite System (GLONASS) of the Russian Federation. Other similar systems are the upcoming European Galileo positioning system, the Japanese Quasi-Zenith Satellite System (QZSS), and the proposed COMPASS-Bediou Navigation System of China.

Reasons which may affect the availability and signal quality of the standalone satellite navigation systems and related algorithms, and consequently quality of clock synchronization, have been extensively discussed in literature e.g., a comprehensive overview can be found in [18, 19]. Special considerations related to the availability of GPS signal refer specifically to mobile CSs. GPS is not available in building or under roofing (e.g., in a wood), which is very likely to (temporarily) happen for mobile CSs. Such mobile CSs that operate on batteries are energy sensitive. Since a GPS sensor is very expensive in terms of consumed energy these mobile applications can benefit by switching off the GPS whenever possible and as long as possible, still maintaining the required synchronization quality.

Additionally, it should be considered that GPS may be subject to deliberate attacks, which even when detected timely they still make the GPS signal unavailable for the whole duration of the attack [5]. Amongst these, we mention [20] (i) jamming GNSS based vehicle tracking devices to prevent a supervisor's knowledge of a driver's movements, or avoiding road user charging; (ii) rebroadcasting ('meaconing') a GNSS signal maliciously, accidentally or to improve reception but causing misreporting of a position; (iii) spoofing GNSS signals to create a controllable misreporting.

In a report from the US Government Accountability Office (GAO) to the US Congress [3] on *"GPS-Disruption—Efforts to Assess Risks to Critical Infrastructure and Coordinate Agency Action Should be Enhanced"* it is pointed out that many of the large infrastructure SoSs in the US are already using GPS time synchronization on a wide scale and a disruption of the GPS signals could have a catastrophic effect on the infrastructure. In this report it is noted that a global notion of time is required in nearly all infrastructure SoSs, such as telecommunication, transportation, energy, etc. and this essential requirement has been met by gradually using more and more often the time signals provided by GPS, not considering what consequences a disruption of the time distribution, either accidental or intentional, has on the overall availability and function of the infrastructure [6].

Even more recently, on 4 February 2016 the BBC reported that "several companies were hit by hours of system warnings after 15 GPS satellites broadcast the wrong time, according to time-monitoring company Chronos" [4]. This led to serious problems to many companies that resulted in money loss.

These events just confirm existing warnings from different communities. For example, a Report from the UK Royal Academy of Engineering in 2011 [20] suggests that the U.K. may have become dangerously over-reliant on satellite-navigation signals, and too many applications have little or no back-up were these signals to go down. The report concludes that several concerns are bounded to GNSS. First, non-GNSS based back-ups are often absent, inadequately exercised or inadequately maintained". Second, that the jammers are easily available and that most jammers are able to block GPS, GLONASS and GALILEO. Third, a full picture of the dependencies on GPS and similar systems is missing.

Starting from the above concern, we can conclude that *we should not entirely rely on satellite navigation systems to build a dependable global time base in time-aware SoS*. Following this observation, this Chapter presents the design and development of a resilient fail-silent master clock based on satellite-based time synchronization (e.g., GPS or Galileo signals), to provide a dependable global time base for cyber-physical Systems-of-Systems. Such Resilient Master Clock (RMC) is intended to feature low power consumption, low weight and low cost. The RMC should be built with hardware

Off-the-Shelf (OTS), as for example COTS MEMS sensors, and whenever possible software OTS.

The RMC includes an independent oscillator and GPS devices complemented by acceptance tests. In fact, software clock control techniques are devised in order to:

1. Provide a self-estimation of the quality of clock synchronization. This is achieved via the Reliable and Self-Aware Clock (R&SAClock), which acts as an oracle of the quality of clock synchronization. R&SAClock keeps nodes of a network aware about the quality of synchronization: it monitors the synchronization level of the local clock with respect to a global time reference (like the Temps Atomique International, TAI).
2. Extend the holdover duration of the clock by compensation of local clock deviations, especially in case of absence of the GPS signals. In fact, clock (crystal oscillators) deviations may be caused by physical environmental variations, like temperature, pressure, humidity variations, voltage. Correction techniques based on COTS sensors are introduced to compensate local clock deviations and avoid unsynchronized clocks in SoSs in the absence of GPS signals.

When the satellite-based time synchronization signal fails (or it is corrupted by a security incident), for a certain amount of time, the RMC is able to maintain its clock close to the global time within a required accuracy until the satellites signal becomes available again.

The rest of the section is organized as follows. Section 2 presents the architecture design of the Resilient Master Clock. The successive Sects. 3 and 4 explore the main technical solutions that are included in the Resilient Master Clock, that are respectively intended to discipline the clock when the GPS signal is unavailable, and to provide a self-estimation on synchronization uncertainty. Section 5 presents conclusions.

2 Resilient Master Clock (RMC) Architecture

This section discusses the architecture of the RMC. The presentation of the architecture is generic i.e., it is not bound to a specific board or pieces of hardware. For example, in Sects. 3 and 4, two different instantiations of (part of) the RMC on two different boards will be described.

The architecture of the RMC is represented in Fig. 1, and described below. In Fig. 1, the components in light grey are required hardware and software components that should be available for the selected OTS board. Dark grey components identify instead the components that have been devised and developed when building the Resilient Master Clock.

The architecture of the RMC is divided in three layers: the *board* (or Hardware layer), the *Operating System* (or OS layer) and the *Middleware*. Each layer consists of the different constituent blocks which are herewith described.

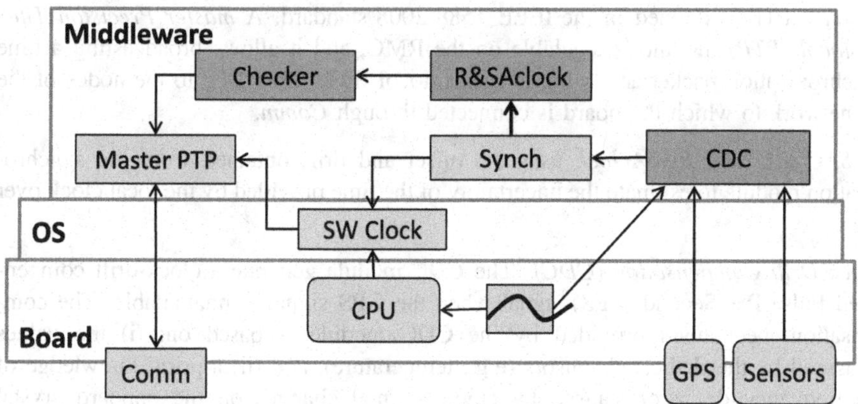

Fig. 1. Resilient Master Clock architecture.

Board Layer. The building blocks of this layer are hardware components. These are:

- A *GPS module* for receiving time messages by the GPS satellite constellation. The messages are then provided to any enquiring hardware or software along with a one-pulse-per-second (1PPS) signal.
- *Sensors*. Sensors for acquiring information about the environment. Examples are temperature and pressure sensors.
- *Comm*. This block refers to the communication interface. For example, an Ethernet Network Interface Card (NIC) can connect the RMC to a network in which CSs slaves wait for time synchronization packets from the RMC, which acts as a master clock node.
- *CPU, Memory and the physical oscillator*. These are standard components of any hardware board. The physical clock is particularly relevant in our context.

OS Layer. This layer includes a local software clock (*SW Clock*) which is usually created by Operating Systems starting from the hardware clock and that provides the timestamps to the services executing on the board.

Middleware Layer. This layer includes OTS SW components (*Synch, master PTP*) and SW components that are specific for the RMC. In particular, these are the following.

Synch. The Network Time Protocol (NTP [17]) is a networking protocol for clock synchronization between computer systems over packet-switched, variable-latency data networks. NTP is intended to synchronize all participating computers to within a few milliseconds of Coordinated Universal Time (UTC). Since RMC is based on TAI, it should be remarked that conversion UTC-TAI is trivial [26]. A *synchronization module* (*Synch*) based on the Network Time Protocol (NTP) uses the GPS time signals to discipline the local clock.

MasterPTP. The Precision Time Protocol (PTP, [7]) is a protocol used to synchronize clocks throughout a computer network. On a local area network, it achieves clock accuracy in the sub-microsecond range, making it suitable for measurement and control systems. PTP is defined in the IEEE 1588-2008 standard. A *master Precision Time Protocol (PTP)* module is available on the RMC, and it allows broadcasting a time synchronization packet according to the protocol IEEE 1588 PTP to the nodes of the subnetwork to which the board is connected through *Comm*.

R&SAClock. The *R&SAclock* uses the offset and drift obtained from the synchronization module to estimate the uncertainty of the time provided by the local clock over time.

Clock Drift Compensation (CDC). The *CDC* module generate a clock-drift compensated Pulse Per Second (PPS) signal when the GPS signal is unavailable. The compensation mechanism provided by the CDC module is based on: (i) the values measured by the dedicated sensors (e.g., temperature), and (ii) a-priori knowledge of the frequency deviation caused by environmental changes on the onboard crystal oscillator (e.g., temperature variations).

Checker. A *checker* module checks the uncertainty associated to the time of the local clock provided by the R&SAclock; consequently, it decides if the RMC can be considered a reliable time source and allows or blocks the PTP synchronization. For example, it can be implemented as a process which periodically checks the quality of the local time provided by the R&SAClock. On the other hand, when the quality of the local time is outside acceptable thresholds, the PTP synchronization beans must be stopped because the RMC cannot be considered a time reference.

3 The Clock Drift Compensation Module

This section discusses the Clock Drift Compensation (CDC) module that provides a periodic Pulse Per Second (PPS) time signal to the Synch module (cf. Fig. 1). During normal operation, the CDC module forwards the high quality, externally provided time signal (e.g., generated by a GPS receiver). In case the externally provided time signal becomes unavailable, the CDC module switches seamlessly without interruption of the output PPS signal to holdover mode until the external time signal becomes available again. During holdover mode the CDC module internally generates the output PPS time signal for the Sync module from a local clock based on a common (quartz) crystal oscillator. This local clock is drift compensated with respect to the – now in holdover mode unavailable – external time signal. The drift compensation improves the precision of the internally generated PPS time signal and consequently allows for a prolonged holdover duration compared to a crystal oscillator based clock that is not clock drift compensated.

In the following subsections we detail our clock drift compensation method and present a proof-of-concept prototype implementation which we used for calibrating and evaluating the CDC module.

3.1 Compensating the Drift of Clocks Based on Crystal Oscillators

Clocks in computer systems are usually realized by a digital counter register and a crystal oscillator whereas each oscillation generates a tick event that increments the counter register. The oscillator frequency output slightly deviates from its designed nominal frequency output, because of (1) mechanical imperfections introduced during manufacturing of the oscillator, (2) dynamic deviations caused by aging of the oscillator, and (3) environmental conditions (e.g., temperature, acceleration, humidity) acting on the oscillator. The static and dynamic deviations from the nominal frequency are the cause for clock drift: Any two clocks of the same design, even when perfectly started at the same instant, will eventually drift apart as time progresses.

For establishing a global time in Cyber-Physical Systems-of-Systems (CPSoSs), we need to periodically resynchronize the local clocks of the Constituent Systems (CSs) by external clock synchronization (e.g., synchronization with the GPS time source). The resynchronization is necessary to ensure a bounded offset, also called precision, among the clocks of the CSs. A critical parameter of clock synchronization is the resynchronization period which needs to be short enough to keep all clocks within the required precision. By actively compensating the clock drift, the duration between two resynchronization instants can be increased. This is of particular interest if the source for external synchronization is unavailable (e.g., losing the GPS signal when driving into a tunnel, turning the GPS receiver off to save power), while the synchronization precision of the clocks of the CSs has to be maintained until the external source becomes available again.

Drift Compensation Method. In order to compensate clock drift, the effects of internal and external sources of oscillator frequency deviations must be negated. This requires to measure these effects or know a-priori about them, but also to apply corrective actions on the clock. There are two options to apply corrections: (1) control the enclosing environment of the oscillator (e.g., oven-controlled or voltage-controlled oscillators) such that the output frequency is corrected, or (2) to periodically correct the counter register by adding each correction period a correction value that compensates for the frequency deviations.

The first approach requires, additionally to the measurement of the enclosing environment, possibly expensive and power-demanding actuation hardware (e.g., heat source), insulation, or is in case of some effects (e.g., acceleration) infeasible. However, for some effects (for example, temperature, humidity, pressure, aging) this approach is technically more simple, as it only requires to steer the oscillator frequency close to the external clock source and then maintain the same environmental conditions during holdover mode.

The clock drift compensation of the CDC module is based on the second correction approach which also depends on measuring the oscillator environment, but – besides that – can be realized purely in software. This software implements a compensation model which predicts for each correction period an accurate correction value in clock ticks. To achieve a high level of accuracy, the clock drift compensation model fuses several sources of information (a-priori knowledge, sensor observations).

Important parameters of our drift compensation method – besides the drift compensation model – are the tick rate of the external time signal which should be forwarded as the output tick by the CDC module during normal operation and generated when absent during holdover mode, the correction period, and the internal tick rate or oscillator frequency. The output tick rate determines the counter register size, i.e., this register needs to be able to count the number of internal ticks that correspond to one output tick. The output tick rate also determines the correction period, because corrections predicted by the compensation model should be applied for each output tick to avoid imprecision effects. Finally, in order to have only a small discretization error, the clock compensation method assumes that the output tick rate (e.g., 1 Hz) is much lower than the internal tick rate (e.g., at least a few kHz or MHz).

Compensation Model. The clock drift compensation model is based on (1) a-priori knowledge about the oscillator (e.g., manufacturing defects, aging behavior, known effects of environmental conditions), and (2) on the currently observed external time signal and the environmental conditions. Environmental conditions that can be observed by sensors are for example: temperature, barometric pressure, acceleration, air humidity.

Explicitly defining a compensation model would be a tremendous task, because it requires precise knowledge about all relevant physical properties of the crystal oscillator, their interrelationships, and the involved sensors which – similarly to the oscillator – also slightly deviate from their designed characteristics (measurement errors that depend on currently prevailing environmental conditions). Consequently, we focus on defining and parameterizing the compensation model by using classical machine learning techniques where, for example, regression (curve fitting) for independent input variables, or artificial neural networks for input variables with unknown dependency relationships are available. The selection of a concrete machine learning technique depends on the available sensors, available computational resources and the required compensation quality. Regardless of the technique, training of the compensation model is necessary by taking a set of input instances (e.g., an observation of the environmental conditions) and adjusting the model parameters such that it maps the input data to a known correction value for the counter register. The trained model is then available during holdover mode for predicting correction values for new input data where the correction value wasn't known before.

For model training there are two methods that should be applied in combination:

- *Offline-Learning/Calibration*: After manufacturing the CDC module, including its oscillator and the sensors, obtains training data by observing (using its own sensors) the controlled environmental conditions. Under these controlled environmental conditions, the oscillator frequency deviation is measured by external equipment (e.g., an oscilloscope that measures the difference of the external time signal with respect to the internally generated output tick during holdover). Training data is collected by doing various sweeps of the controlled environmental conditions through the range of expected environmental conditions and recording the sensor observations together with the necessary correction value. Offline learning initializes the compensation model with a-priori knowledge.

- *Online-Learning/Adaptation*: During operation of the CDC module the compensation model can be adjusted by constantly retraining it, when the external time signal is available. Online-learning compensates for aging effects of the oscillator and involved sensors. Also online-learning possibly allows for limiting the necessary offline-learning to a smaller sample size of a production series where only small model deviations are expected among individual CDC modules.

3.2 Proof-of-Concept Prototype

The proof-of-concept prototype is based on an implementation of the CDC module on the SmartAP 2.0[1], a small embedded system originally intended for auto-piloting small aircrafts. It consists of a STM32 ARM Cortex M4 microcontroller integrated with two external quartz oscillators (84 MHz, and 32.768 kHz), and sensors to measure acceleration (Invensense MPU-6050/ MPU-9150), barometric pressure (MS5611-01BA03). We customized the board by adding sensors to additionally measure temperature and air humidity (Sensirion SHT75). As an external time signal we used an UBLOX LEA-6H GPS receiver. For wirelessly communicating with the board we added a Bluetooth module (Microchip RN42).

The microcontroller implements a data recording functionality to obtain the measured training data, and the drift compensation method including a simple variant of a compensation model. In this prototype we did not implement online-learning, because the effects of online-learning are minor, if the compensation model is well calibrated and the CDC module prototype has not been left several months for aging between calibration and evaluation.

Compensation Model Implementation. Figure 2 illustrates a simple compensation model based on look-up tables where averaged training data can be deployed directly. For each environmental condition a look-up table exists from which the contribution of the measured parameter to the clock drift is obtained. Temperature is codependent with all other environmental conditions. Consequently, the lookup tables for pressure, humidity, and acceleration contain temperature dependent correction values. To estimate the clock drift for the current correction period, the individual contributions are summed up and added to the constant drift value of the crystal oscillator.

Training of the Compensation Model. The compensation model training data is obtained by placing the CDC module with its sensors in an experimental chamber within which the environmental conditions can be controlled. For each correction period an oscilloscope records the deviation of the internally generated output time signal from the external time signal (GPS receiver).

Figure 3 shows our prototype of such an experimental chamber. It consists of a thermally insulated box with a Peltier element for heating and cooling, an air pump to produce an over or under pressure inside, and a humidifier. To achieve a constant acceleration force on the oscillator crystal, the board is mounted on a plate that can be

[1] http://sky-drones.com/autopilots/9-smartap-autopilot-20.html.

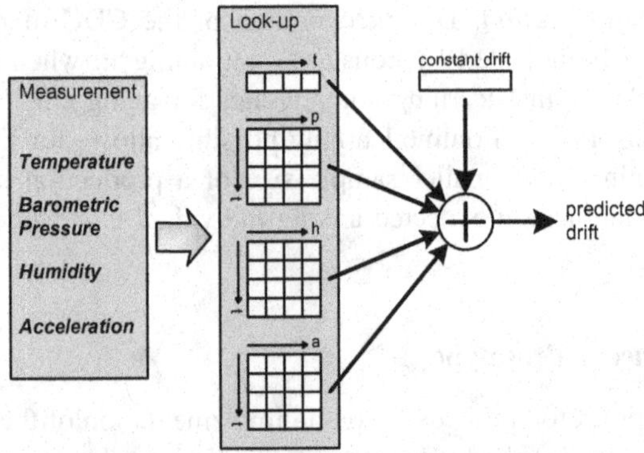

Fig. 2. Clock drift compensation model based on look-up tables.

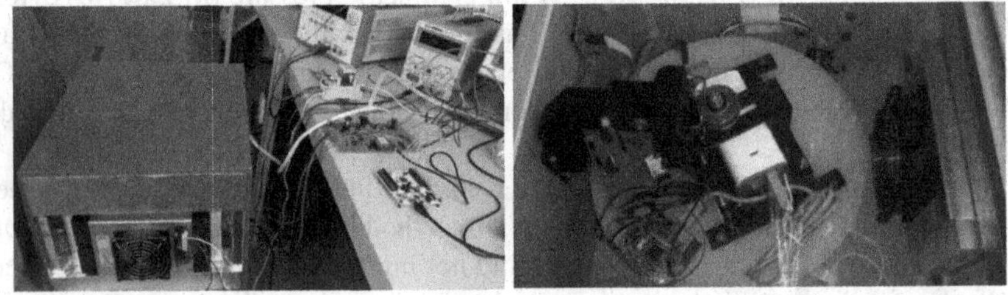

Fig. 3. Experimental chamber for obtaining training data for the compensation model. On the right the prototype of the CDC module has been highlighted with a red circle and is mounted on the rotating plate.

rotated. Acceleration forces can be applied on the oscillator in X and Y directions, depending on how the board is mounted on the rotating plate.

By using this box, one environmental condition after the other is varied from its minimum value up to the maximum, and vice versa. For each set point the number of oscillations of the crystal oscillator, which drives the clock, is counted several times. The reading and resetting of this counter is triggered every second by the PPS time pulse originating from the GPS receiver.

From the recorded training data, the constant drift at certain environmental conditions – i.e., the zero conditions C_0 – is obtained. C_0 can be selected arbitrarily. Furthermore, for each condition that can be observed a table entry (see lookup-based compensation model in the previous section) is derived by averaging over the training data that have been recorded for the same environmental conditions. These table entries indicate the additional drift if the conditions deviate from the zero conditions C_0.

3.3 Evaluation

The evaluation of the CDC module is conducted using the abovementioned experimental chamber. In particular, only results for temperature and acceleration in X

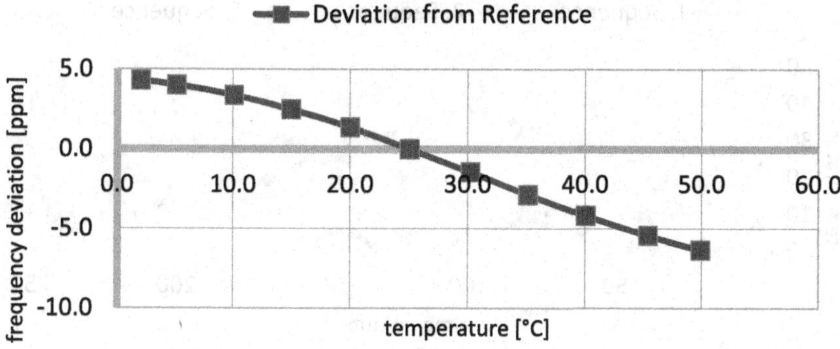

Fig. 4. Frequency deviation with respect to temperature.

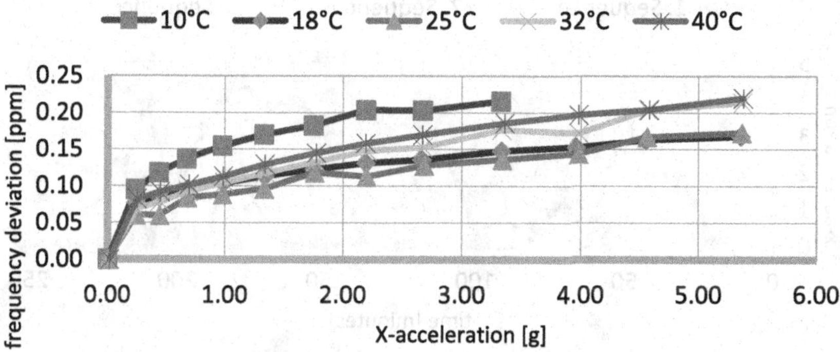

Fig. 5. Frequency deviation with respect to acceleration in X-axis at different temperatures.

direction for the 84 MHz quartz oscillator are shown here, while information also on other environmental conditions (Y-acceleration, air humidity, barometric pressure) and the second oscillator can be found in [15].

Obtaining Training Data. The investigated oscillator has a measured variance (i.e., variations in the number of oscillations when environmental conditions are stable) around its nominal frequency of approximately 0.02381 ppm. In the following we present frequency deviations of the oscillator when doing a temperature sweep from 0° C to 50° C and an acceleration sweep from 0 g to 5 g (g is the weight per mass unit).

Figure 4 depicts the frequency deviation over the temperature range. If, for instance, the frequency deviation is about 2.5 ppm (e.g., at 15° C), it means that in each second a clock driven by this oscillator deviates 2.5 μs from a clock with perfect accuracy. Without clock synchronization, this deviation sums up to 9 ms/h, or 0.216 s/day. In Fig. 4, a positive value of the deviation denotes an increase in the number of oscillations compared to the nominal value.

Figure 5 shows the frequency deviation of the oscillator concerning variations in the acceleration force and temperature.

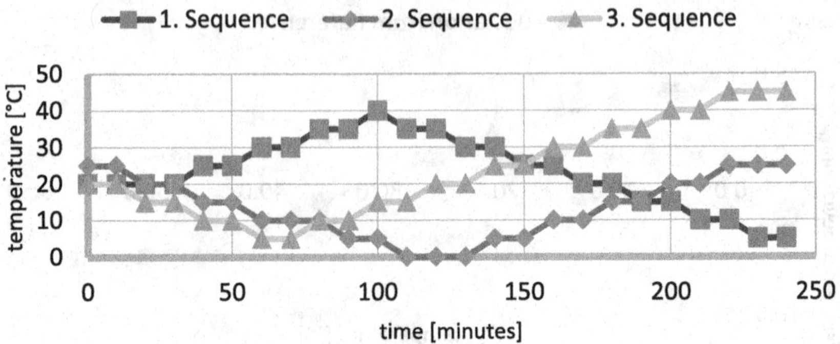

Fig. 6. Three test sequences of temperature set points.

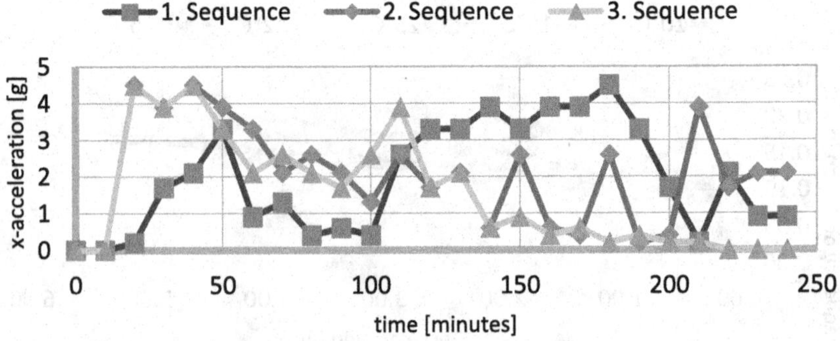

Fig. 7. Three test sequences of X-acceleration set points.

The results show that additionally to the significant correlation between the temperature and the frequency deviation, the oscillator exhibited deviations concerning changes in acceleration (and pressure and humidity, see [15]). All these deviations have been higher than the variance of the oscillator at stable environmental conditions.

Evaluation of Compensation Model. The evaluation of the compensation model carried out by recording the compensation performance over a set of test sequences where the controllable environmental conditions are varied. Each of the test sequences takes 240 min. Figures 6 and 7 show three of such sequences for the environmental conditions temperature and acceleration in X direction.

Figure 8 shows the compensation performance of the CDC module in holdover mode only for the first test sequence (1. Sequence), because all other test sequences gave similar results. In Fig. 8 we plotted the results for two different compensation models: basic compensation (only constant drift correction), and temperature, acceleration and pressure (T&A&P). Also for comparison reasons, the deviation is depicted, when no compensation mechanism is applied. Different effects during manufacturing or aging of the oscillator crystal lead to a permanent deviation from the nominal frequency (here it is faster than the reference), which is about 60 ppm.

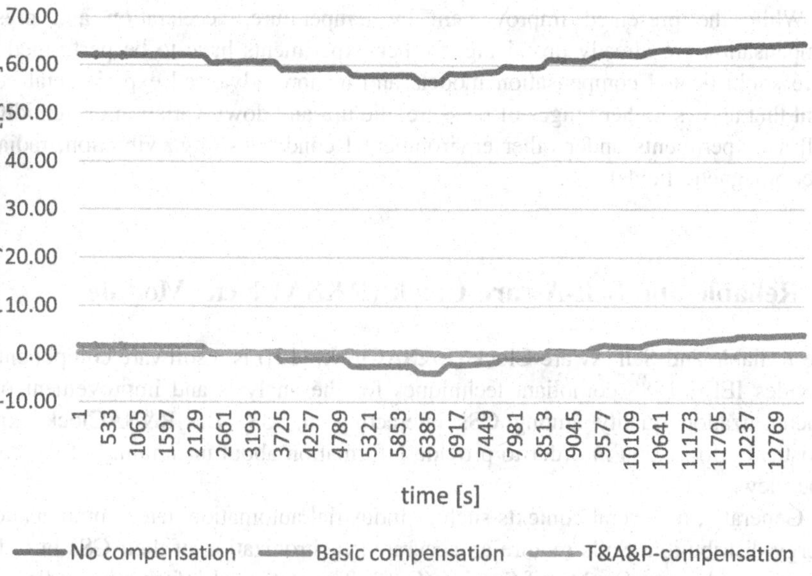

Fig. 8. Frequency deviation under different compensation models.

Clearly, basic compensation has the largest corrective effect and improved the mean drift rate of the oscillator from an order of magnitude of about 10^{-5} to approximately 10^{-6}. The compensation model that regarded more of the investigated environmental conditions (T&A&P) gave a better result, as the oscillator's drift rate is now in the 10^{-7} order of magnitude.

Even more improved results should be easily achievable when using more sophisticated compensation models, sensors of better quality, and implementing online-learning for self-fine-tuning. Consequently, this evaluation gives strong support to the benefits of our proposed clock drift compensation method.

3.4 Summary of Main Findings on the Clock Drift Compensation Module

The proof-of-concept prototype confirms that the compensation of frequency deviations of crystal oscillators by passive observation of the surrounding environmental conditions, and using a trained compensation model leads to a significant decrease of clock drift. The different environmental conditions indeed have an effect on the stability of the oscillator and some of these effects can be reduced when these conditions are known. Available protocols (e.g., the Network Time Protocol – NTP) are able to compensate constant drifts of the local clock of a computer system, if the environmental conditions are not changed after the reference clock is disconnected. However, many CPSoSs operate in environments, within which it is infeasible – or at least only with a considerable effort (e.g., by constantly heating the crystal oscillator) – to keep these conditions stable. In contrast, sensors to determine the environmental conditions are often already available in those systems, or can be installed at relatively low cost.

While the presented improvement by temperature, acceleration and pressure compensation are already promising, further experiments have to be performed with more sophisticated compensation models and a more advanced experimental equipment that allows higher ranges of pressure, the up- and down variation of humidity, as well as experiments under other environmental conditions (e.g., vibration, radiation, electromagnetic fields).

4 Reliable and Self-Aware Clock (R&SAClock) Module

The Reliable and Self Aware Clock (R&SAClock, [12]) is a software component that provides IEEE 1588 compliant techniques for the analysis and improvement of the synchronization quality among CSs interacting in SoS. The R&SAClock exploits statistical information in order to provide information about uncertainty of the current time view.

Generally, in several contexts such as industrial automation, telecommunication or energy distribution, SoSs require an accurate synchronization of their CSs in order to assure the adequate Quality of Service (QoS). The statistical information collected by R&SAClock to estimate synchronization uncertainty [8] is used as feedback about quality of synchronization. The CSs equipped with R&SAClock are continuously updated about the current synchronization performance.

4.1 General Concepts on R&SAClock

A CS uses R&SAClock to acquire both the time value and synchronization uncertainty associated with the time value.

For clarity, we report basic notions on time and clocks that are used in the rest of this section. Noteworthy, the terminology is consistent with the terms defined in Chap. 1. Figure 9 below is introduced to better clarify relevant aspects.

The *global time* is an abstraction of physical time in a distributed computer system; it is the unique time view shared by the CSs. The *reference clock* is a *working hypothesis* for measuring the instant of occurrence of an event of interest: it is a clock that always holds the global time. We can say that the *reference node* is the CS that owns the reference clock. Also, given a local clock c and any *time instant* t, we define $c(t)$ as the *time value* read by local clock c at time t.

The behavior of a local clock c is characterized by the quantities *offset*, *accuracy* and *drift*. The *offset* of two events denotes the duration between two events and the position of the second event with respect to the first event on the timeline; the offset $\Theta_c(t) = t - c(t)$ is the actual distance of local clock c of the CS node n from the global time at time. This distance may vary through time.

Accuracy A_c of clock c denotes the maximum offset of a given clock from the external time reference, measured by the reference clock. An upper bound of the offset adopted in the definition of system requirements and therefore targeted by clock synchronization mechanisms.

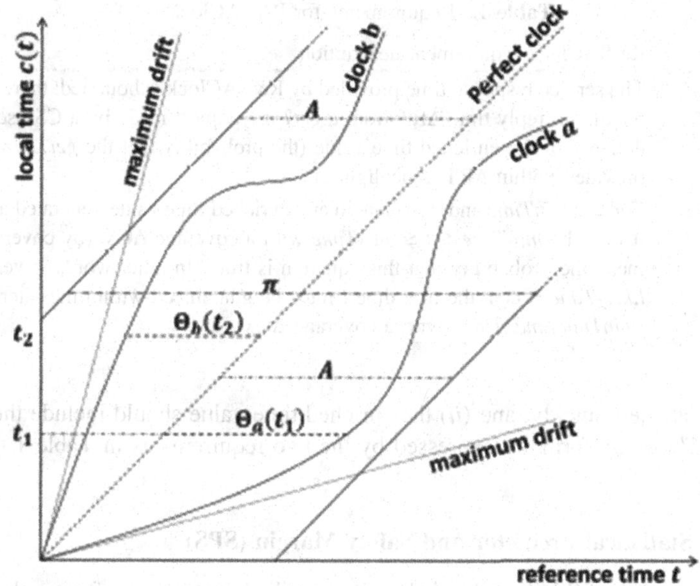

Fig. 9. Basic notions on time and clocks.

The precision π of an ensemble of synchronized clocks denotes the maximum offset of (distance) respective ticks of the global time of any two clocks of the considered clock ensemble.

The drift $\rho_c(t)$ of a physical clock describes the frequency ratio between the physical clock and the reference clock i.e., the rate of deviation of a local clock *c* at time *t* from global time [10].

Synchronization uncertainty $U_c(t)$ is defined as an adaptive and conservative evaluation of the offset $\Theta_c(t)$ at any time *t*; uncertainty is such that $A_c \geq U_c(t) \geq |\Theta_c(t)| \geq 0$ [8]. Hence, accuracy Ac is an upper bound of uncertainty $U_c(t)$ and consequently of the absolute value of the offset $\Theta_c(t)$.

When a CS asks the current time to R&SAClock, the latter provides an enriched time value useful for time synchronization. The enriched time value is composed of a set of values: *likelyTime*, *minTime*, *maxTime* and *FLAG*. *LikelyTime* is the time value computed reading the local clock. *minTime* and *maxTime* represent left and right synchronization uncertainty margins with respect to *likelyTime*. They are based on synchronization uncertainty provided by the internal mechanisms of R&SAClock. Finally, the *FLAG* takes the value 1 if requirements on uncertainty are satisfied, 0 otherwise. Details on R&SAClock and its implementation can be found in [8, 11].

It is evident that the main core of R&SAClock is the uncertainty evaluation algorithm that equips R&SAClock with the ability to compute the uncertainty. Such an algorithm relies on the Statistical Predictor and Safety Margin (SPS) algorithm.

Each CS that uses the R&SAClock *getTime* method for getting synchronization information and each CS has the two main expectations: (*i*) a request for the time value

Table 1. Requirements for R&SAClock

Req. ID	R&SAClock requirement description
REquation 1	The service response time provided by R&SAClock is bounded: there exists a maximum reply time ΔRT from a *getTime* request made by a CS user to the delivery of the enriched time value (the probability that the *getTime* is not provided within ΔRT is negligible)
REquation 2	For any *minTime* and *maxTime* in any enriched time value generated at time t, it must be *minTime* ≤ t ≤ *maxTime* with a coverage ΔCV (by coverage we mean the probability that this equation is true). In other words, given *likelyTime* = c(t), the true time t must be guaranteed within the interval [*minTime*, *maxTime*] with a coverage ΔCV

should be satisfied quickly, and (*ii*) the enriched time value should include the correct real time. These are formally expressed by the two requirements in Table 1.

4.2 The Statistical Predictor and Safety Margin (SPS)

In the following the SPS algorithm is briefly described for a local software clock *c* that is disciplined by an external clock synchronization mechanism. SPS computes the uncertainty at a time *t* with a coverage, intended as the probability that $A_c \geq U_c(t) \geq |\Theta_c(t)| \geq 0$ holds. The computed uncertainty is composed by three quantities: (i) the *estimated offset*, (ii) the output of a *predictor* function, *P* and (iii) the output of a *safety margin* function, *SM*. The computation of synchronization uncertainty requires a right uncertainty $U_r(t)$ and a left uncertainty $U_l(t)$: consequently, SPS has a right predictor with a right safety margin for right uncertainty, and a left predictor with a left safety margin for left uncertainty. The output of the SPS at $t \geq t_0$ is constituted by the two values:

$$U_r(t) = \max(0,\ \tilde{\Theta}(t_0)) + P_r(t) + SM_r(t_0) \tag{1}$$

$$U_l(t) = \min(0,\ \tilde{\Theta}(t_0)) + P_l(t) + SM_l(t_0) \tag{2}$$

The *estimated offset* $\tilde{\Theta}(t_0)$ is computed by the synchronization mechanism and can contain errors. If the estimated offset is positive, it influences the computation of an upper bound on the offset itself and consequently is considered in (1). If it is negative, it is ignored. A symmetric reasoning holds for (2).

The *predictor* functions, $P_r(t)$ and $P_l(t)$, predict the behavior of the oscillator and continuously provide bounds (lower and upper) which constitute a safe (pessimistic) estimation of the oscillator drift and consequently a bound on the offset. The oscillator drift is modelled with the random walk frequency noise model, one of the five canonical models used to model oscillators (the power-law models [14]), that we considered as appropriate and used. Obviously the parameters of this random walk are unknown and depend on the specific oscillator used. They are computed resorting to the observation of the last *m* samples of the drift (where *m* smaller or equal to the set-up

Table 2. SPS parameters

Symbol	Definition
t_0	time in which the most recent synchronization is performed
$\tilde{\Theta}(t_0)$	estimated offset at time t_0
$\tilde{\rho}(t_0)$	estimated drift at time t_0
M, m	maximum and current number of (most recent) samples of the estimated drift that the UEA collects ($0 < m \leq M$)
N, n	maximum and current number of (most recent) samples of the estimated offset that the UEA collects ($0 < n \leq N$)
p_{ds}	probability that the population variance of the estimated drift is smaller than a safe bound on such variance
p_{dv}	a safe bound of the drift variation since t_0 is computed with probability p_{dv}
$p_{ds} \circ p_{dv}$	the joint probability of these two values represents the coverage of the prediction function
p_{os}	probability that the population variance of the estimated offset is smaller than a safe bound on the variance
p_{ov}	a safe bound of the offset at t_0 is computed with probability p_{ov}
$p_{os} \circ p_{ov}$	the joint probability of these two values represents the coverage of the safety margin function

parameter M), and using a safe bound on the population variance of the estimated drift values. The coverage of this safe bound depends on the set-up probabilities p_{ds} and p_{dv} defined in Table 2 together other main quantities involved in the SPS algorithm.

The *safety margin* functions $SM_r(t_0)$ and $SM_l(t_0)$ aim at compensating possible errors in the prediction or in the offset estimation. The safety margin function is computed starting from the collection of the last n samples of the estimated offset (where n is smaller or equal to the set-up parameter N). A safe bound to the population variance of the estimated offset is computed. The coverage of this safe bound depends on the set-up probabilities p_{os} and p_{ov} (see Table 2).

The parameter t_0 is the time in which the most recent synchronization is performed. At time t_0 the synchronization mechanism computes the estimated offset $\tilde{\Theta}(t_0)$ and possibly the estimated drift $\tilde{\rho}(t_0)$ (if not provided by the mechanism, it can be easily computed by R&SAClock itself).

4.3 Proof-of-Concept and Exemplary Runs

The R&SAClock has been implemented in the Beagle Bone [16] board with Debian OS as described in the AMADEOS deliverable D4.4 [15]. The R&SAClock uses GPS for clock synchronization; it acquires data on the estimated offset and drift from the Network Time Protocol (NTP) [9] component.

When the synchronization uncertainty exceeds a given threshold, the Checker module is notified by reading the FLAG field of the enriched time value. The communication channel between the Checker and the R&SAClock is socket-based.

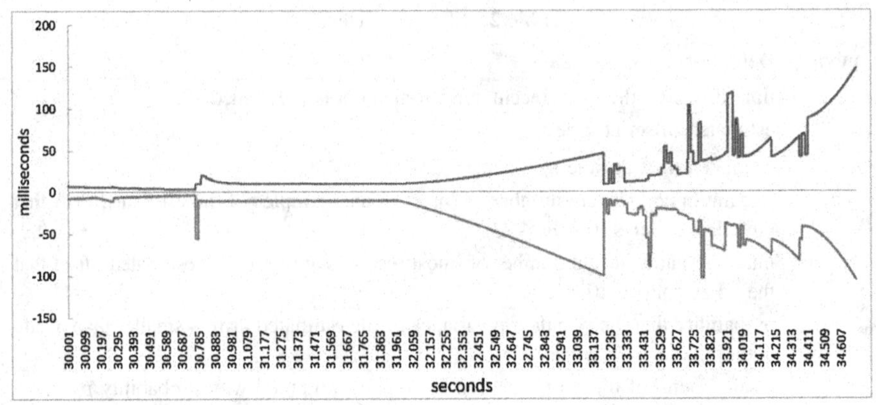

Fig. 10. Exemplary run of R&SAClock on the proof-of-concept.

In the following we show an exemplary run with the R&SAClock executing on the Beagle Bone board proof-of-concept. We acknowledge that an extensive assessment activity is required [13] (*i*) to give evidence that the defined software executing on the proof-of-concepts satisfies the identified requirements, and (*ii*) to opportunely tune the R&SAClock parameters (Table 2). Still, we present this exemplary run because it explains intuitively the behavior of R&SAClock.

In the considered run, reported in Fig. 10, the *x*-axis (in seconds) corresponds to the *likelyTime* collected reading the local clock. The *maxTime* and *minTime* computed by R&SAClock are respectively the two lines above and below the *x*-axis. In fact, the *y*-axis (in milliseconds) shows the *maxTime* and *minTime* with respect to *likelyTime* i.e., *maxTime-likelyTime* and *likelyTime - minTime*. The *FLAG* value is not shown in this figure.

In the present run, the R&SAClock was already running for 30000 s (see x-axis). As long as the connection with the GPS signal is stable, NTP reliably disciplines the local clock: an accurate estimation of offset is provided, and synchronization uncertainty is a small interval (in the order of few microseconds or less). As it can be shown in the time interval between second 30000 and 30600, the synchronization uncertainty is slightly reduced through time. In fact, the R&SAClock "studies" the past behavior of the local clock, understand that it is overall stable and trustable, and reduces synchronization uncertainty.

Instead, at approximately second 30700, an instability in the local clock and NTP is detected, most likely due to the temporary unavailability of the GPS signal (signal loss). Synchronization uncertainty is increased, because no fresh information on the clock behavior w.r.t. the reference time is provided.

At approximately second 31900, the synchronization uncertainty steadily increases through time. In fact, GPS signal is lost and no fresh information from the time source is provided. There are no guarantees that the clock is disciplined correctly.

When the GPS signal is newly available (approximately at second 33200), a new accurate estimation of the offset is provided. Consequently, the synchronization uncertainty is reduced again. However, from now on, the synchronization with the GPS is unstable: there are only few, sparse synchronizations. This determines the behavior of the R&SAClock, which cannot trust the local clock and consequently the synchronization uncertainty grows.

4.4 Summary of Main Findings on R&SAClock

The R&SAClock was initially proposed in [8, 11, 12] to monitor the software clock in distributed system. In such works, R&SAClock was implemented and exercised on a fixed node, and with the intention of supporting only the node itself.

Instead, in the RMC, the R&SAClock is intended to operate as a failure detector for a Master clock: in other words, the Checker module of the RMC can read the FLAG value of the R&SAClock, and decide if the RMC can act as a master clock or not.

In addition, we implemented the R&SAClock on a light board which has small requirements for power consumption. This improves the range of applicability of the R&SAClock w.r.t. the previous environments described in [12], which are distributed servers. The experiments, although still preliminary, confirm that the R&SAClock behaves as expected also confirming, in a different environment, the results shown in [12].

5 Conclusions

This Chapter discussed the role of time in Systems-of-Systems. Building on the terms, definition and knowledge defined in Chap. 1, this Chapter identified motivation, with examples, for the prominent role of time and clocks in time-aware Systems-of-Systems. Further, the Chapter discussed the challenges of resilient time keeping and it presents the Resilient Master Clock (RMC), a hardware-software solution that acts as an accurate, fail-silent global time base which is externally synchronized to a satellite-based time source.

The design of the RMC is presented, its main algorithm illustrated including results from the execution on two different prototypes. Although significant work is still needed to consolidate results, the RMC appears a promising approach to provide a low-cost, low-power consumption solution for resilient time-keeping and resilient master clock in Systems-of-Systems.

References

1. Kopetz, H.: Why a Global Time is Needed in a Dependable SoS. CoRR abs/1404.6772 (2014)
2. AMADEOS Consortium, D2.3 – AMADEOS conceptual model – Revised (2016)
3. US Government Accountability Office: "GPS Disruptions: Efforts to Assess Risk to Critical Infrastructure and Coordinate Agency Actions Should be Enhanced". Washington, GAO - 14-15 (2013)

4. GPS error caused ‘12 hours of problems’ for companies, 4 February 2016. http://www.bbc.com/news/technology-35491962
5. Shepard, D.P., Humphreys, T.E., Fansler, A.A.: Evaluation of the vulnerability of phasor measurement units to GPS spoofing attacks. Int. J. Crit. Infrastruct. Prot. **5**(3), 146–153 (2012)
6. Cyber-physical Systems-of-Systems: the AMADEOS approach and Main Advances, Webinar for the INCOSE WG on Systems-of-Systems (2015)
7. 1588-2008 - IEEE Standard for a Precision Clock Synchronization Protocol for Networked Measurement and Control Systems (2008)
8. Bondavalli, A., Ceccarelli, A., Falai, L.: Assuring resilient time synchronization. In: Proceedings of the 2008 Symposium on Reliable Distributed Systems, October 06–08, 2008, SRDS, pp. 3–12. IEEE Computer Society, Washington, DC
9. IEEE Standard for a Precision Clock Synchronization Protocol for Network Measurement and Control Systems, IEEE Std 1588-2008 (IEEE Std 1588-2002), p. cl-269, 24 July 2008
10. Verissimo, P., Rodriguez, L.: Distributed Systems for System Architects. Kluwer Academic Publisher, Dordrecht (2001)
11. Bondavalli, A., Brancati, F., Ceccarelli, A.: Safe estimation of time uncertainty of local clocks. In: Proceedings of International IEEE Symposium on Precision Clock Synchronization for Measurement, Control and Communication, ISPCS 2009, pp. 47–52
12. Bondavalli, A., Brancati, F., Ceccarelli, A., Falai, L., Vadursi, M.: Resilient estimation of synchronisation uncertainty through software clocks. Int. J. Crit. Comput. Based Syst. **4**, 301–322 (2013). doi:10.1504/IJCCBS.2013.059038. ISSN: 1757-8779
13. Bondavalli, A., Brancati, F., Ceccarelli, A., Vadursi, M.: Experimental validation of a synchronization uncertainty-aware software clock. In: The 29th IEEE Symposium on Reliable Distributed Systems, Delhi, India, October 31 – November 3, 2010, pp. 245–254 (2010). doi:10.1109/SRDS.2010.35. ISBN: 978-0-7695-4250-8
14. Barnes, J.A., et al.: Characterization of frequency stablity. IEEE Trans. Instrum. Meas. **IM-20**, 105–120 (1970)
15. AMADEOS project, deliverable D4.4 “Internal Delivery- Design of Resilient Master Clock” (2016)
16. BeagleBoard.org project Debian. http://beagleboard.org/project/debian
17. Mills, D.: Internet time synchronization: the network time protocol. IEEE Trans. Commun. **39**(10), 1482–1493 (1991)
18. Beekhuizen, J., Kromhout, H., Huss, A., Vermeulen, R.: Performance of GPS devices for environmental exposure assessment. J. Exposure Sci. Environ. Epidemiol. **23**(5), 498–505 (2013). ISSN 1559-0631
19. Kaplan, E.D., Hegarty, C.J. (eds.): Understanding GPS: Principles and Applications, 2nd edn. Artech House, Boston (2006)
20. The Royal Academy of Engineering, Global Navigation Space Systems: reliance and vulnerabilities, March 2011
21. Lamport, L.: Time, clocks, and the ordering of events in a distributed system. Commun. ACM **21**(7), 558–565 (1978)
22. Verissimo, P.: On the role of time in distributed systems. In: FTDCS (1997)
23. Verissimo, P., Rodrigues, L.: Distributed Systems for System Architects, vol. 1. Springer Science & Business Media, New York (2012)
24. Ceccarelli, A., et al.: Introducing meta-requirements for describing system of systems. In: 2015 IEEE 16th International Symposium on High Assurance Systems Engineering. IEEE (2015)

25. Kopetz, H.: Real-time Systems-Design Principles for Distributed Embedded Applications. Springer, New York (2011)
26. Levine, J., Mills, D.: Using the network time protocol (ntp) to transmit international atomic time (tai). In: 2nd Annual Precise Time and Time Interval Systems and Applications Meeting, 28–30, Reston, VA, USA, pp. 431–440 (2000)

Managing Dynamicity in SoS

Sara Bouchenak[1(✉)], Francesco Brancati[2], Andrea Ceccarelli[2], Sorin Iacob[3], Nicolas Marchand[1], Bogdan Robu[1], and Patrick De Oude[3]

[1] Université Grenoble Alpes, Grenoble, France
sara.bouchenak@insa-lyon.fr,
nicolas.marchand@gipsa-lab.fr,
bogdan.robu@gipsa-lab.grenoble-inp.fr
[2] Department of Mathematics and Informatics, University of Florence, Florence, Italy
francesco.brancati@resiltech.com,
andrea.ceccarelli@inifi.it
[3] Thales Netherlands B.V., Hengelo, The Netherlands
{sorin.iacob,patrick.deoude}@nl.thalesgroup.com

1 Introduction

SoS dynamicity refers to short-term changes in an SoS, which occur in response to changing environmental or operational parameters of the CSs. These changes may have different effects, such as SoS adaptation or the generation of emergent phenomena. This chapter starts by recalling the MAPE approach in Sect. 2 before to introduce existing monitoring approaches in Sect. 3. Finally, Sect. 4 overviews existing reconfiguration techniques for SoS dynamicity management, related to Analyzis, Planning and Execution phases and illustrates through an examples the possible implementations of dynamicity management with modelling and feedback control techniques.

2 Overall MAPE Approach

We follow the classical MAPE-K control loop for designing an autonomic manager over managed elements (Fig. 1). This consists mainly in components to Monitor, Analyze, Plan and Execute the reconfiguration plan. When an SLA (Service Level Agreement) and its associated service level objectives are associated with the service of a managed element, the MAPE control loop guarantees that those service level objectives are met, and if it is not the case, a new plan is calculated and used to reconfigure the system.

This work has been partially supported by the FP7-610535-AMADEOS project.

A. Bondavalli et al. (Eds.): Cyber-Physical Systems of Systems, LNCS 10099, pp. 186–206, 2016.
DOI: 10.1007/978-3-319-47590-5_7

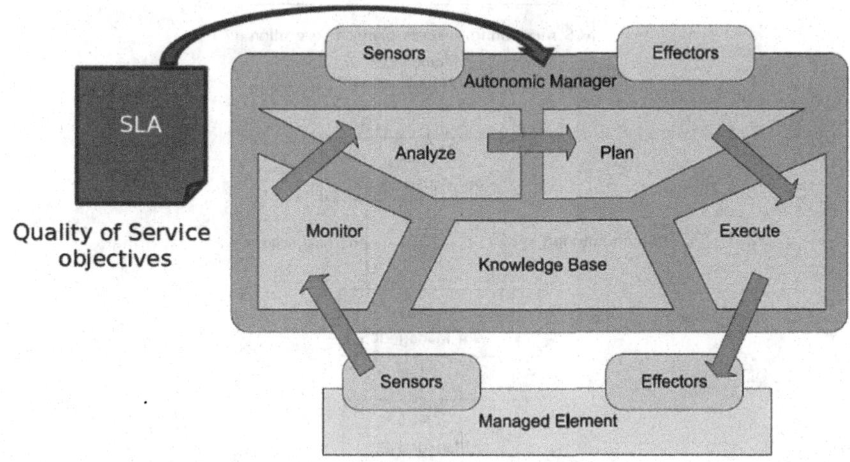

Fig. 1. MAPE control loop

2.1 SoS Management Infrastructure

In this section we present how to exploit the AMADEOS SysML profile described in Chap. 4 in order to build the infrastructure for MAPE purposes. To this end, we implemented through the profile six MAPE architectural patterns. Thus, for each of the six patterns, we realize a SysML block diagram built using the stereotypes defined in the profile.

Each CS is represented as a block and it has an interface, either a RUMI or a RUPI interface in order to enable the exchange of messages and physical entities respectively. Each CS may implement the SoS management activities, namely monitoring, analysis, planning and execution.

In the following we report the (1) Hierarchical Control, (2) Master/Slave, (3) Regional Planner, (4) Coordinated control, (5) Information sharing, (6) Atomic patterns and a conclusive analysis on their recursive inclusion in a SoS.

2.2 Hierarchical Control Pattern

In the hierarchical control pattern, we have a managed CS which is controlled directly by a managing CS by means of their RUMI interfaces over which monitoring information and the enacting actions are transmitted. As we can notice (Fig. 2), the managing CS has been stereotyped with all the functions of the MAPE cycle. We present below the UML representation of such pattern.

2.3 Master/Slave Pattern

In the master/slave pattern, a set of managing CSs shares part of the MAPE functions (Fig. 3). In our instantiation, master CS (Managing_AP_CS) performs analysis and planning activities and two slave CSs perform monitoring and execute functions

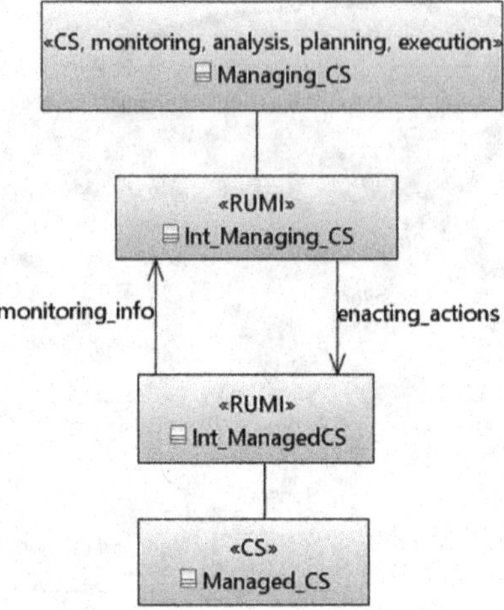

Fig. 2. Hierarchical control pattern

(Managing_ME_CSx). The slave CSs send monitored_info to the master and they receive planned info from the master CS. This information is exchanged through the RUMI interfaces of slave CSs and the master CS. Finally, the slave CSs are in charge of communicating each with a managed CS. The latter transmit monitored data to the slave CS, which in turn forwards back the enacting actions as planned by the master CS.

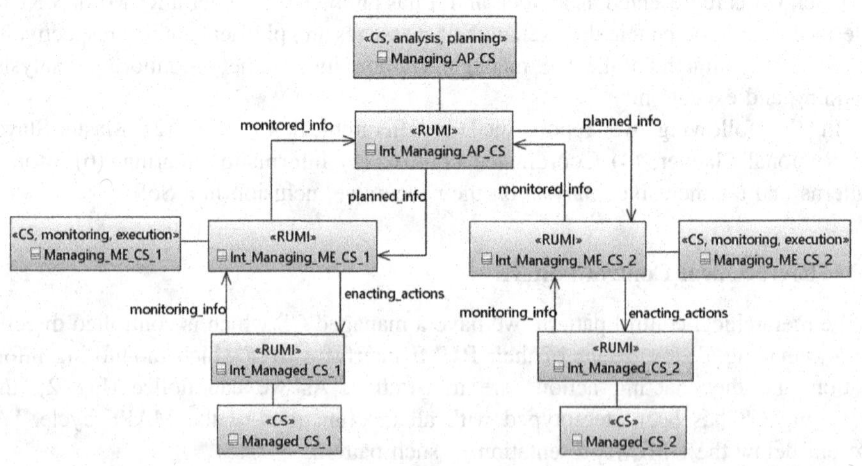

Fig. 3. Master-slave pattern

2.4 Regional Planner Pattern

In the Regional Planner, a regional managing CS (Managing_P_CS_x) implements only the Planning activity (Fig. 4). Instead, Analysis, Monitoring and Execution are

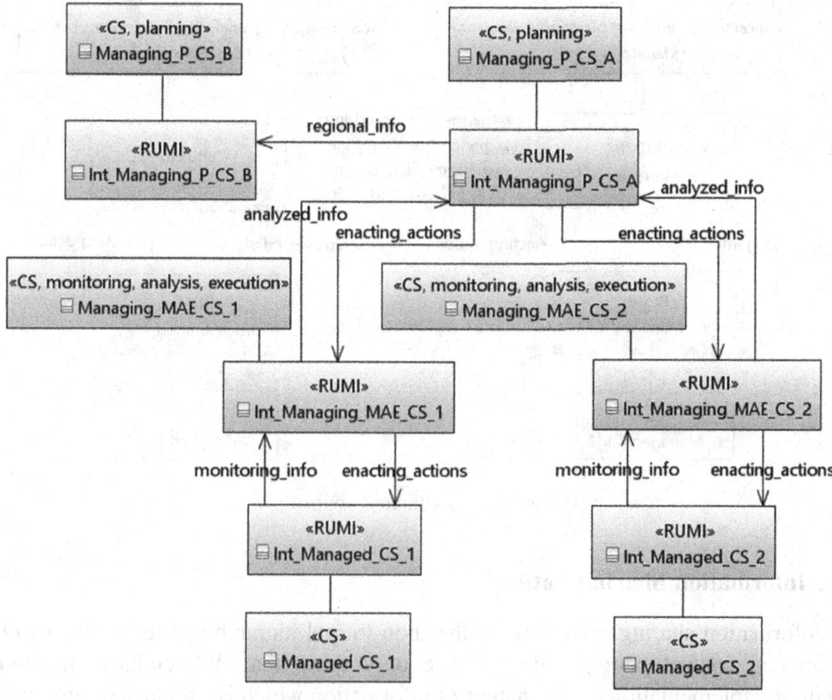

Fig. 4. Regional planner

delegated to other managing CSs (Managing_MAE_CS_x). The regional managing CS is responsible for a region of CSs and it exchanges with other peer CSs the information on a regional basis. At the bottom level, each managed CSs (Managed_CS_x) send monitoring information to its corresponding managing CS (Managing_MAE_CS_x), which performs analysis activities and then forwards the results to the regional CS (Managing_P_CS). The latter performs the Planning and forward enacting actions back to the managing CS. Finally, the managing CS enacts such actions towards the managed CSs.

2.5 Coordinated Control Planner

In the following, we present the implementation for the non-formal hierarchy pattern Coordinated Control pattern (Fig. 5). The coordinated control pattern consists of a set of CS implementing all the MAPE phases (Managing_CS_x) and exchanging through their RUMIs the monitored, analysis, planned and execution information. Through RUMIs, the managing CSs can collect monitoring info from the managed CS (Managed_CS_x) and forward the actions to enact.

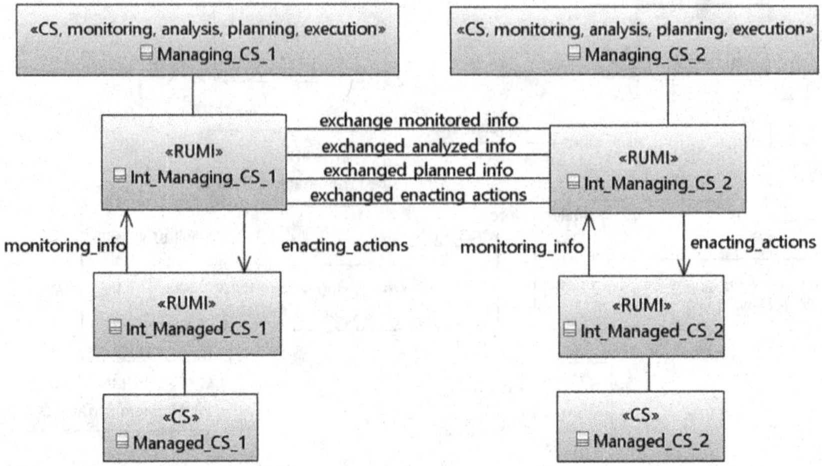

Fig. 5. Coordinated control

2.6 Information Sharing Pattern

The information sharing pattern is another non-formal hierarchy pattern similar to the coordinated control pattern (Fig. 6). The only thing that differentiates the corresponding implementations is the nature of information which is exchanged through the RUMIs of the managing CSs. In the sharing information pattern only monitored data are exchanged while the rest of information is not shared among managing CSs (Managing_CS_x).

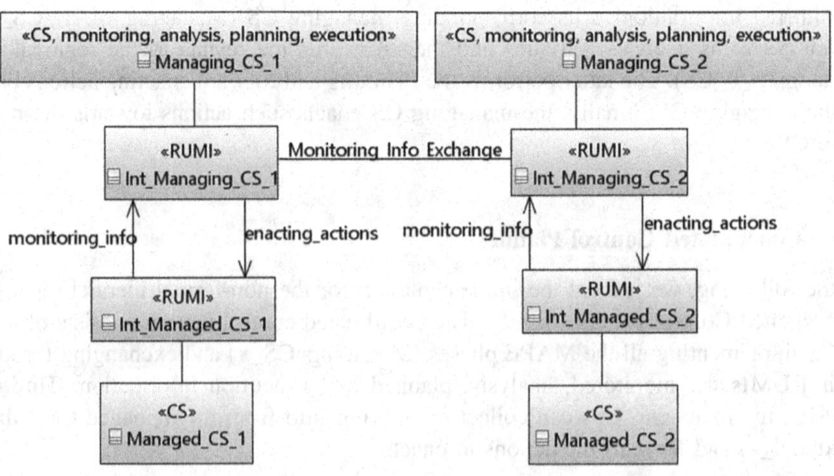

Fig. 6. Information sharing

2.7 Atomic Pattern

In the following, we present the atomic pattern to enable the interaction of a CS with the physical environment (Fig. 7). To this end, the managing CS (Atomic_CS) carries out the MAPE cycle once it has received monitoring information through its RUPI interface and it forwards physical signal over the same interface. Physical entities received through the RUPI interface come from the afferent environment while the outgoing flow of physical entities is forwarded to the efferent environment.

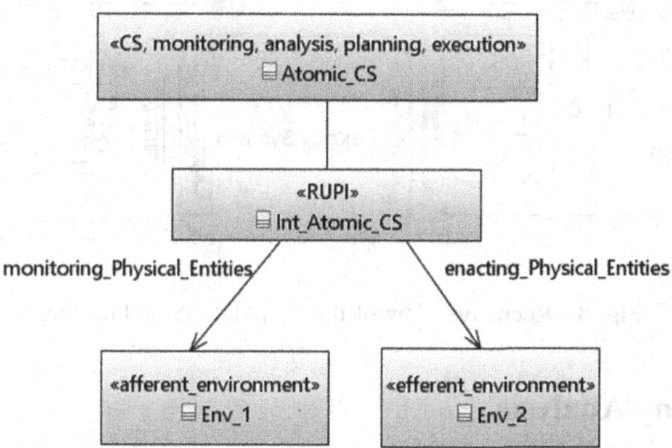

Fig. 7. Atomic pattern

2.8 Recursive Inclusion of Patterns for SoS Design

At SoS level, the MAPE should be applied recursively. Each CS implementing part of the MAPE may in turn be substituted recursively by another pattern which is realized through a further set of CSs. This approach foresees the possibility to have several nested levels of MAPE, each of them belonging to a different hierarchical level of the SoS, or holarchical set of CSs. Distinct set of CSs are observed by a MAPE in which the Adaptive Monitoring, the Analyzer and Planner, and the Reaction Strategies coexist and cooperate.

Let us consider the scenario depicted in Fig. 8, in which several CS are represented and controlled by an implementation of the MAPE-K cycle in which we have a dedicated CS for Monitoring, another for Analysis and Planning and a third one for Execution.

The Adaptive Monitoring (M) is able to detect events according to the QoS specifications of the CS. The Cognitive and Predictive Models (AP) search for the causes and the effects correlating data incoming from the Adaptive Monitoring. The Reaction Strategies (E) perform the proper recovery action. The considered SoS can be integrated in a more complex SoS; Fig. 8 gives a representation of the recursive foreseen architecture.

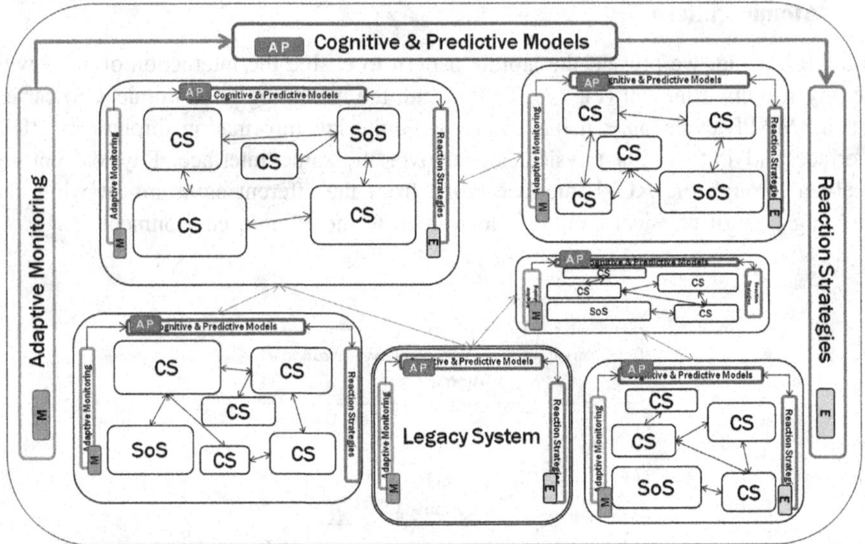

Fig. 8. Recursive view of the AMADEOS architecture

3 Monitoring/Analysis

In this section, we report on the broad topic of monitoring in SoSs. In particular, focusing on observation and data analysis (i.e., the Monitoring and Analysis components of the MAPE building block). We investigate basics on monitoring and detection (Sect. 3.1) and main monitoring approaches (Sect. 3.2).

3.1 Basics on Monitoring and Detection in SoSs

It is very important to guarantee that an SoS behaves as expected. To this end, monitoring activities control the SoS by means of verifying that system behavior and performance comply with well-defined rules. Verification activities can be carried out at two different stages either on-line, i.e., while monitoring data are collected, or off-line i.e., after the collection process.

In monitoring literature, the system which has to be monitored is called target system while the hardware component and the software application within the target systems are called respectively target component and target application. In our case a CS represents the target system while the target component and the target application are represented by the physical and software part of the CS itself. We refer to the CS which receives monitored information as monitoring CS.

Monitoring activities consists in observing behavior and performances of CS target components in order to collect useful information to guarantee the correct SoS functioning. Monitoring activities by themselves are not sufficient to guarantee the correct behavior of an SoS, but they have to be integrated with techniques to diagnose the SoS behavior in its execution environment. To this end, we have presented in deliverable D3.1 [1] an SoS Management Infrastructure which complements Monitoring with Analysis, Planning and Execution (MAPE) facilities. This section focuses on the Monitoring (M), and partially also on the Detection of events (Analysis of data), consequently addressing the MA letters of the MAPE loop. These two actions (MA) are in fact often tightly bounded, because a monitoring system is usually conceived and instantiated with the specific intention of detecting specific events or verifying that certain conditions are met.

Let us now focus on the way monitoring activities are carried out. Essential objects that are exploited to monitor the system behavior are the so called probes. Probes can be inserted either inside or outside the target system and provide useful information on how the system behaves. As an example Fig. 9 shows the probes inserted within the target system which can also provide the system intermediate output (see Fig. 9-b) and it shows the possibility of monitoring the system as a black box (see Fig. 9-a).

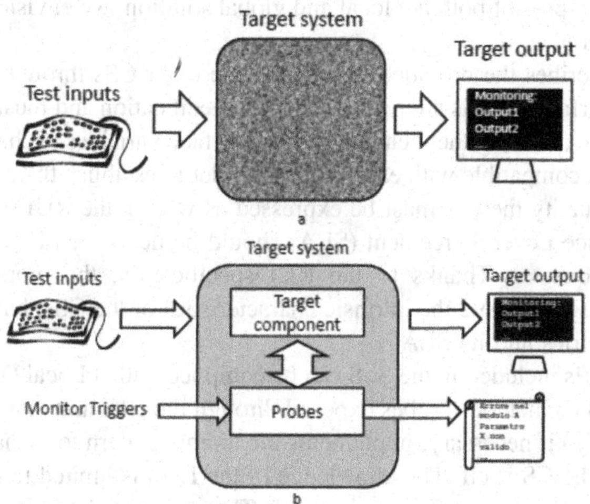

Fig. 9. Black box (a) and instrumented (b) monitoring of the target system

Probes can be hardware or software. In the first case hardware signals are monitored, while in the second case, code is inserted within the target application to collect internal information of the system (code instrumentation). Two rules have been defined which have to be respected by probes:

- they should observe as much information as necessary to satisfy the objectives of the monitoring activities,
- they should not compromise, or at least compromise as little as possible, the behavior of the target system.

3.2 A General Approach to SoS Monitoring

Considering the complexity and heterogeneity of an SoS, it is necessary that a monitoring solution deals with the issue of where to deploy the monitoring and detection system. It could be placed locally on each CS or globally at a higher level. Both approaches have advantages and disadvantages with respect to the CS and/or the SoS, as listed in the following:

- local solution pros: allows to perform a more precise detection activity on the single CS due to the perfect knowledge of CS itself.
- local solution cons: could negatively affect the performance of each CS, thus compromising the overall performance of the SoS.
- global solution pros: allows to improve detection accuracy, like detection of detrimental emergence phenomena.
- global solution cons: requires a large amount of data to be transferred from each CS, thus potentially affecting the network bandwidth in negative way.

To capture the pros of both the local and global solution, we envision the following overall architecture.

Each CS describes the provided services to the other CSs through the RUI specification. The interface models are part of the RUI specification and must be based on an agreed ontology explaining the meaning of the interface variables exchanged across the RUI and must be compatible with each other. In order to establish the desired quality of service (QoS), quality metrics must be expressed as well in the RUI specification. For example, a Service Level Agreement (SLA) should be negotiated between the service provider and requester. Thanks to the RUI specification, the monitoring and the detection systems can ignore the intrinsic characteristics of the CSs, but they are aware of its quality metrics and its SLA.

Each CS that is included in the SoS can be equipped with a Local Detection System (LDS). The LDS (i) includes probes exposed through the RUI and that are necessary to observe events; (ii) if necessary, implements the atomic pattern to manage the physical environment of the CS itself. The knowledge of the LDS is limited to the CS: in other words, the other connected CSs are ignored. The different detectors, relying on the exposed probes, can be organized and coordinate following the different MAPE patterns.

At SoS level, a global detection system will be also deployed. It differs from the ones deployed at a local level because it has an overall view of the SoS and consequently it has the ability to observe and detect events as a combination of the outputs of the individual LDSs. Furthermore, it may consider different and additional quality metrics and indicators with respect to the LDSs. The global detection system fetches data from the LDS or MAPE instantiations available in the SoS, and perform global monitoring and analysis, acting according to the master/slave pattern.

Figure 10 shows a high level representation of the architecture of the envisioned system. Specifically, the CSs composing the SoS are represented in different shapes, to show that they are different one from the other. Each CS is able to communicate with the others by means of the RUI interface, represented by the box labelled RUI. Finally, the local detection systems, labelled LDS are also included in the representation of each CS, in order to eventually detect anomalous events on the corresponding CS. At the top

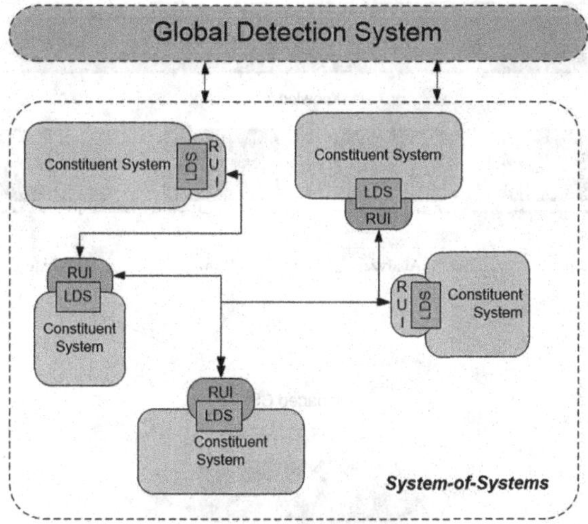

Fig. 10. Monitoring infrastructure in SoSs

of the figure, the Global Detection System is also shown, which observes the status and the events that are happening at the SoS level.

Depending on the monitoring purposes, confidentiality and privacy issues may need to be guaranteed through proper data security and anonymization. Especially, while anonymization solutions can be executed locally, confidentiality requires that the endpoint agrees on adequate secure communication protocols.

4 Analysis/Planning/Execution

4.1 Overview

The main challenges related to the dynamic adaptation of CPSoS stem from the distributed nature of the measurement and control infrastructure (MAPE).

Since the Monitoring and Analysis blocks have been discussed earlier, this section focuses primarily on a potential design of the Planning and Execution blocks.

The M and A functions of the MAPE building block determine the values of the CPSoS parameters, whereas the P and E functions close the control loop either by generating control signals for the CS, or by adjusting the environmental parameters of the CS, as depicted in Fig. 11.

All MAPE functions are instantiated for a particular CPSoS dynamicity model, which specifies which state parameters need to be measured, what metrics have to be used for combining or aggregating these parameters, how the control must be implemented to achieve the desired effect, and how this control algorithm generates CS control parameters. Obviously, this domain knowledge applies in a similar way for all the composite MAPE patterns described in the AMADEOS deliverable D3.1 [1].

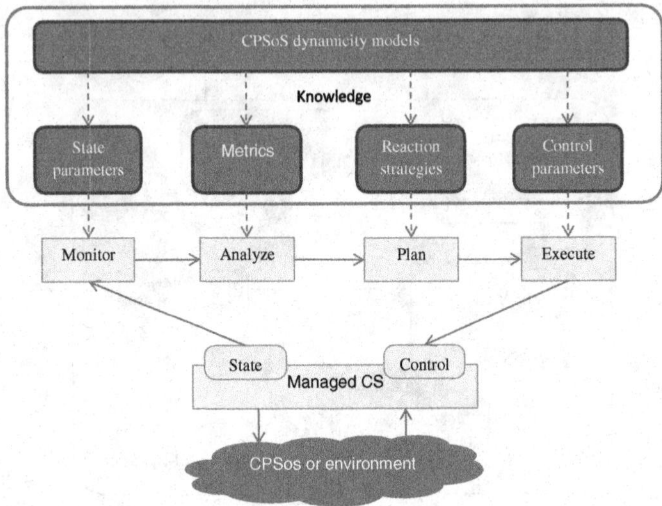

Fig. 11. Close loop control involving a CS and a MAPE block

To emphasize the control aspect, we redraw the MAPE control loop as in Fig. 12. Although not fundamentally different from a traditional control loop, Fig. 13 emphasizes the potential difference between the input and output parameters ($Y(t)$ and $X(t)$, respectively) toward the CPSoS and the monitoring and control parameters ($\mu(t)$, and ε(t), respectively) toward the MAPE blocks. This is useful for CPS, since the monitoring and control of the RUPI-based interactions can also be executed through RUMI, which adds flexibility to the definition of reaction strategies. The (⊕) blocks denote some suitable transformations that map the external and control input vectors onto a common metric space, and the ⊕ block combines these values in a single metric to obtain the control value. Based on this value the Control block generates a vector $\sigma(\mathrm{t})$

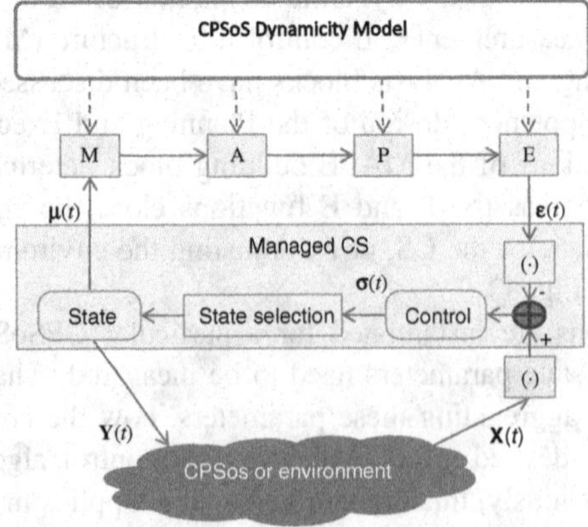

Fig. 12. Control and feedback parameters in MAPE-based control loops

whose elements show the mismatch between the current and desirable state parameters, so a new state can be selected that compensates for this mismatch.

4.2 Example: Control Loop for Electrical Vehicle System

If the CS is a Charging Point (see AMADEOS use case [2, 3]) Y(t) in Fig. 13 can be a state vector specifying the charging current, maximum allowed power phasor variations, and last kWh price. When an Electric Vehicle (EV) is connected to the CP, the State includes, among other functions, a metering function which could generate the monitoring vector $\mu(t)$ as a sequence of messages at regular time intervals, containing instantaneous measurements of the current drawn by the EV, and the maximum available current for that CP. Similarly, X(t) could be the remaining charging time as estimated by the EV, and $\varepsilon(t)$ could be a sequence of asynchronous (i.e. event-triggered) messages containing a new value for the kWh price, and a new maximum power rating for the CP.

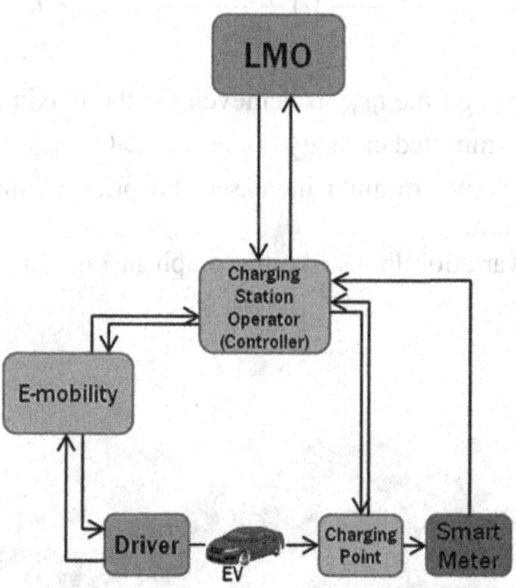

Fig. 13. Simplified functional diagram for an electrical vehicle charging station

Obviously, the summation block in Fig. 12 assumes an appropriate abstraction of the two inputs (MAPE controls and external CPSoS inputs), which is achieved through some mappings. In the example considered above both the remaining charging time and the new kWh price could be mapped on some real value expressing the economic efficiency of the CP.

To illustrate several reaction strategies, we construct a simple example derived from the AMADEOS EV Charging use case. Assume that a Charging Station includes *N* charging Points (see Fig. 13), has a total charging capacity Q, and a maximum charging rate I¬max. Each of the CP has a maximum charging rate ICP_{maxICP}, with $NI\_CP_{max} = aI_{max}$, and a > 1, which means that when all CPs are in use, not all of them can deliver the maximum charging rate. To compensate for this limitation, the CP can influence the charging demand by increasing the energy unit prices. In general, the energy unit price has to be agreed between the EV user and the CSO before the charging operation starts, and should not be changed until the charging is complete. For this reason, it is convenient to define a charging contract in terms of total requested electric charge by user i, Q_i, and the maximum charging duration t_i. For simplicity, we assume that the minimum charging times only depends on the maximum current that a CP outlet can provide.

If the CSO chooses to guarantee a constant charging rate I_i, then the charging time for EV i will be constant as well (within some uncertainty limits): $t_i = \frac{Q_i}{I_i}$. Although it is both in the interest of the CSO and the user to minimize t_i, this is not achievable simultaneously for all the CPs.

A good pricing strategy should try to approach the maximum charging current of the CSO(I_{max}), without lowering the charge unit price below a given minimum. As an example one could consider the following pricing function by the CSO, for user i:

$$p_{CSOi}(Q_i, t_i) = p_0 \frac{t_{max} - t_{imin}}{t_i - t_{imin}} \left(1 + \frac{1}{I_{max} - I}\right), t_{min} < t_i \leq t_{max}, I < I_{max} \quad (1)$$

The minimum charging time t_{ijmin} is achieved for the maximum available charging rate given the already committed capacity (I) for the CSO : $t_{imin} = \frac{Q_i}{I_{max} - I}$. The third term of the product in the above formula increases the price as the committed capacity approached the maximum.

Overall, the price variation looks like the graph in Fig. 14.

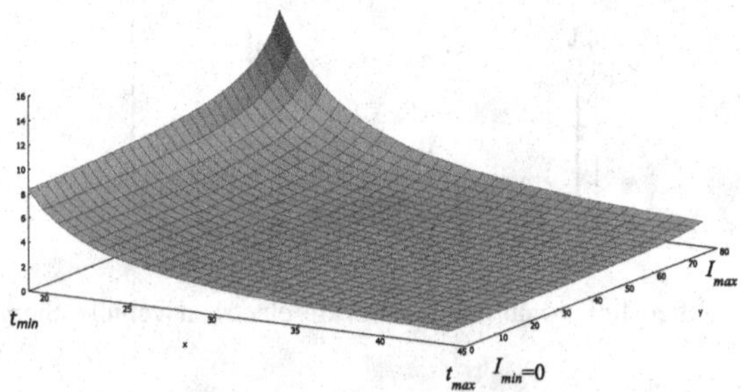

Fig. 14. Possible price adaptation as a function of requested charging time and available

At the same time, a user will always choose a shorter charging time, provided that the price does not increase beyond a predefined personal limit p_{imax}:

$$p_i = \min(p_{imax}, p_{CSOi})$$

Of course, this is a naïve view, since I_j varies in time as other running charging actions end (see Fig. 15), so a significant part of the charging capacity is not used. The

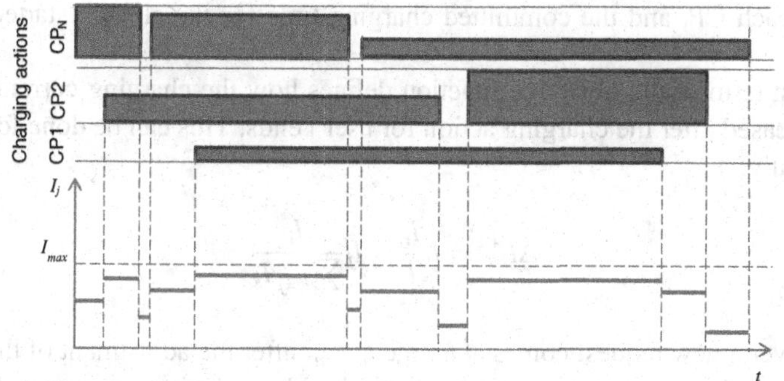

Fig. 15. Variation of the total charging current at CSO

total revenue for the CSO at any given time $P(t) = \sum_{i=1}^{N} p_i(t)$. At the same time, the instantaneous cost for the CSO is given by a component proportional to the total current absorbed plus some constant cost value c_0: $C(t) = c_0 + cI(t)$. The profit made by the CSO can thus be expressed as:

$$\rho_{CSO}(t) = P(t) - C(t) \tag{2}$$

By allowing a variable maximum charging current, the actual charging time results shorter than the one calculated with the pricing model for constant charging current used in this example. The statistics of charging times and new requests will require pricing strategies that take into account the expected variations in the occupancy of CPs.

4.3 Analytic Approaches

When the CPSoS behaves according to a known model expressed analytically, the control can be defined in terms of this model. The pricing model in Eq. (1) allows the CSO to set a dynamic price for a charging operation at constant current. However, this model leads to unused capacity, so it may be advantageous for the CSO to deliver a faster charge if there is unused capacity and other requests are expected to arrive soon.

An optimal control problem with infinite time horizon attempts to maximize some cost or benefit function, such as the one in Eq. (2) [4]. A potential objective function for the optimal control problem could attempt the simultaneous optimisation of the following aspects:

- minimizing the agreed charging time for the already started charging operations.
- maximizing the price for the new charging contract.

The control parameters are the individual charge unit prices and the charging currents for each user. The constraints are the total current for the CSO, the maximum current of each CP, and the committed charging time for the already started charging actions.

The first term of the objective function defines how the charging current for user I can be increased after the charging action for user j ends. This can be done for instance, proportionally:

$$\Delta I_i^+ = I_j \frac{I_i}{I} = I_j \frac{I_i}{\sum_{k \neq j} I_k} \tag{3}$$

Whenever a new request comes at a time t_{i_start} after the adjustment of the charging rates, the new charging current must be maximised by reducing the current for the still running charge operations, down to the limit that would still allow the completion of the charging within the remaining time according to the original contract for user *i*:

$$\Delta I_i^- = \frac{Q_i - \int_{t_{i\_start}}^{t_{j\_start}} I_i(t)dt}{t_{i\_end} - t_{j\_start}} \tag{4}$$

where t_{i_start} and t_{i_end} are the starting and ending times of the charging operation for user i.

Equations (1) to (4) can be used for defining different control approaches, such as Optimal (Stochastic) Control with infinite time horizon [4], Linear Quadratic Controls [5], etc.

4.4 Machine Learning Approaches

In case when the dynamic behaviour of the CPSoS is unknown, or cannot be expressed analytically, data-driven techniques can be used for obtaining an implicit "encoding" of system behaviour. Such an encoding should be able to predict the outputs the system will generate for a given input and contextual parameters.

Machine learning includes a set of techniques that use observations of a system's behaviour patterns to attempt predicting its future states. If we consider the predictive or supervised machine learning approach the goal is to learn a mapping from input values **X** to an output value *y* based on a set of input-output pairs called a training set. The mapping can be represented as a function f_Θ, with a set of parameters Θ, that can be used, given an unseen before pattern $\mathbf{X}_i$, to predict $\hat{y}_i$, i.e. $\hat{y}_i = f_\Theta(\mathbf{X}_i)$. The function f_Θ is what we call a model that is parametrized with a set of parameters Θ. f_Θ can be, in fact, a simple linear function (regression), but generally is a complex algorithm that maps inputs to output values. The goal of machine learning is to find the set of parameters Θ of a chosen model based on a training data set D using a learning algorithm. If the parameters Θ are properly learned then we will be able to estimate the class label $\hat{y}_i$ based on certain unseen input values xj $\notin$ D, where $\hat{y}_j = y_j$ in the majority of the cases.

Artificial Neural Networks (ANN) implement a machine learning technique inspired by a simplified model of the working of the brain. The elementary operations in an ANN are weighted summations of the inputs to obtain an output value. Different output values are obtained by summing the same input with different weights. This way, an input vector is mapped to an output vector. A multi-layer NN combines usually two or more such mapping units. The training of an ANN attempts to adjust the weights such that a particular output is consistently obtained for different input patterns belonging to the same class. When this is achieved, the ANN is said to be able to generalize. Another aspect of training attempts to adjust the weights such that different output patterns are generated for input vectors belonging to different classes. When this is achieved, the ANN is said to be able to discriminate. Training algorithms have been developed that achieve a good trade-off between these two properties. One such algorithm is called the back-propagation learning algorithm that uses an iterative scheme, such as gradient descent [6], to optimize Θ in the learning equation described earlier.

Returning to the EV charging example we want to estimate the best price for a new user i by learning a model based on the following input variables:

- the number of charging points N;
- the requested electrical charge Q_i by user i;
- the maximum charging duration t_{imax} requested by user i;
- the maximum charging durations of the other users j(j$\neq$i);
- the current charging rates I_j of other EVs;
- the maximum charging rate for the CPs;
- the maximum charging rate of the CSO;
- the minimum charging time of EV i, t_{imin};
- the expected number of users in the coming hour (based on historic data);

Let's denote these input variables with x_n. Based on x_n we want to learn a model from which the best price $\hat{p}_i$ can be estimated for a new user i. In order to learn such a model the following cost function will be used

$$J(\theta) = \frac{1}{2}\sum_{n=1}^{N}(p_{ni} - \hat{p}_{ni})^2 \qquad (5)$$

Where $\hat{p}_{ni}$ is the estimated price by the model, p_{ni} is the price associated to the input variables, N is the total number of training samples and n corresponds to the nth training sample $(x_n, p_{ni}) \in D$. In case an ANN is used as a model to learn the best price the gradient descent back-propagation algorithm can be used as an iterative scheme to optimize the model based on the cost function J(θ). By using this iterative scheme, the parameters (i.e., the weights and bias term of each neuron in the ANN) are updated, after applying the training set D, such that the cost function is minimized. When the change of the parameter values are small enough that it can be assumed that the parameters have converged and the model is learned.

The learned solution may not be optimal, since the back-propagation algorithm can be trapped in a local minimum. This is due to the high nonlinear nature of the cost function in the parameter space. Often better performance can be obtained by using a pattern-by-pattern mode, also online mode, to learning the model. In this case the weights of the ANN are updated at every time instance a new pattern is presented.

Additionally, it is also recommended to randomize the data sequence prior to using it for learning. In practice it was shown that the pattern-by-pattern mode result in faster convergence and better solutions [7].

4.5 Feedback Control Approach

In the following, we consider a SoS in which one of the CSs is a computing cluster used to run compute-intensive and/or data-intensive business logic of the cyber-physical system. In the following, we first illustrate the impact of environmental changes and system configuration parameters on the performance and availability of such CSs. We then present a possible implementation of the Analysis component of MAPE through behavioural modelling and a possible implementation of the Planning/Execution components of MAPE through feedback control.

MapReduce is a popular programming model and execution environment for developing and executing distributed data-intensive and compute-intensive applications [8]. However, the complexity of configuration of such systems is continuously increasing. Although the framework hides the complexities of parallelism from the users, deploying an efficient MapReduce implementation poses multiple challenges. MapReduce's ad-hoc configuration and provisioning require a high level of expertise to tune [9]. Ensuring performance and dependability of MapReduce systems still poses several challenges.

One of the most popular open source implementations of the MapReduce programming model is Hadoop. It is composed of the Hadoop kernel, the Hadoop Distributed Filesystem (HDFS) and the MapReduce engine. Hadoop's HDFS and MapReduce components originally derived from Google's MapReduce and Google's File System initial papers. HDFS provides the reliable distributed data storage and the MapReduce engine provides the framework to efficiently analyse this data.

In the following, we consider a CS that is a MapReduce cluster that consists of sub-CSs represented by N nodes. A MapReduce workload is defined as the number of concurrent clients (C) that are sending requests to the central controller. Admission control is a classical technique to prevent server thrashing. It consists of limiting the maximum number of clients (MC) that are allowed to concurrently send requests to the central controller.

The performance of MapReduce systems can be measured as the average time (Rt) needed to process a request in a certain time window. Low client response time is a desirable as it reflects a reactive system. The average *Rt* can, for instance, be calculated at every 30 s, using a sliding window with period 15 min.

$$Rt[s] = avg(Rt_1, Rt_2, \ldots, Rt_N) \tag{6}$$

Availability (Av) refers to the accessibility of the system to users. MapReduce is available if the user requests are accepted at the time of their submission. Availability is instantaneous and concentrates on the fraction of time where the system is operational in the sense of being accessible to the end user. Availability is measured as the ratio of accepted MapReduce client requests to the total number of requests, during a period of time. T here is the previously defined sliding time window size that is used to assign a measurable dynamics to the system. Since T is constant for all experiments, we use only the percentage (%) symbol as the availability measurement unit in all the plots to simplify their understanding.

$$Av\left[\frac{\%}{T}\right] = \frac{N_{SuccessfulJobs}}{N_{SucessfulJobs} + N_{RejectedJobs}} * 100 \tag{7}$$

Furthermore, the service cost is a linear function of the MapReduce cluster size (N), and can be inferred directly from N.

Finally, performance and availability metrics are part of the SLA of the MapReduce system. The SLA specifies MapReduce service level objectives (SLOs) in terms of, for instance, the maximum response time Rt_{max}, and the minimum availability Av_{min} to be guaranteed by the MapReduce system.

Figures 16, 17, 18 show the impact of the variation of, respectively, the workload exogenous variable, the MapReduce cluster size control variable, and the MapReduce cluster's admission control variable on performance and availability metrics. Thus, there is no one-fits-all configuration; rather, a solution that meets a combination of service level objectives as described below.

Example of a Behavioral Model. Capturing the complex behaviour of MapReduce CSs is highly challenging. We propose a model that captures the dynamics of MapReduce CSs, and renders their levels of performance and availability. The model is

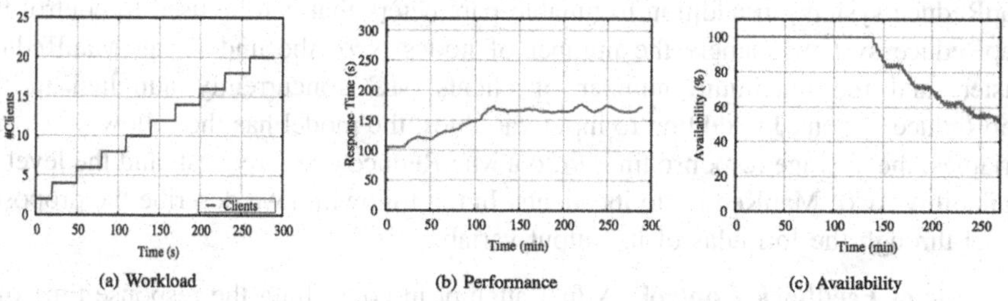

Fig. 16. Impact of workload on MapReduce performance and availability with #Nodes = 20, #MC = 10

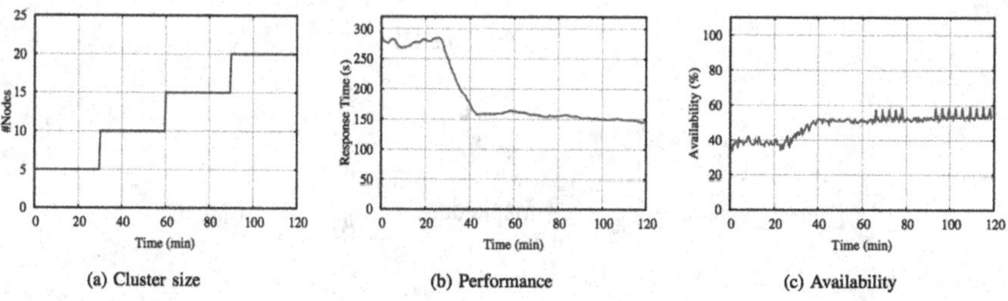

Fig. 17. Impact of cluster size on MapReduce performance and availability with #Clients = 10, #MC = 5

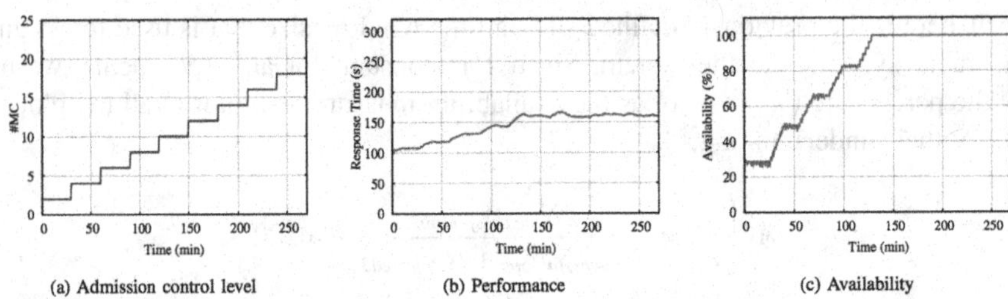

Fig. 18. Impact of admission control on MapReduce performance and availability with #Nodes = 20, #Clients = 10

built as a set of difference equations - as for biological or economical systems - that describe the impact of input variables' variations on system's output variables. We apply a novel modelling approach that considers the MapReduce system as unknown and derives a mathematical model based only on the impact of the input variations on the system's outputs. This technique is part of what we call system identification in control theory. Roughly speaking, one provides known input variation functions (e.g. a step or sinusoidal variation) to the system, and measures the system response to this excitation. Using the output measurements an identification algorithm can approximate the system's internal dynamics. In most cases, without a loss in generality, 1st or 2nd order polynomial difference equations capture the system behaviour sufficiently well.

Figure 19 describes the proposed model variables. The inputs of the model are: exogenous input *C* that represents the number of clients accessing the underlying MapReduce system, in addition to tunable parameters that can be used to control the MapReduce system, namely the number of nodes N of the underlying MapReduce cluster, and the maximum number of clients MC concurrently admitted in the MapReduce system. In addition to input variables, the model has the following output variables: the average response time *Rt* to a MapReduce client request, and the level of availability *Av* of MapReduce to its clients. In the following, we describe the proposed model through the formulas of its output variables.

Example of Feedback Control. A first attempt in controlling the response time of a MapReduce system by adding and removing nodes was realized in [10] by using a PI and a feedforward controller. We design MR – Ctrl, an optimal controller, able to deal

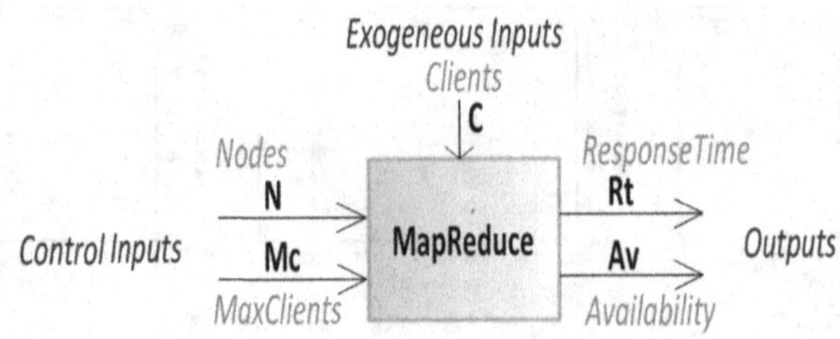

Fig. 19. System model inputs and outputs

with contradictory objectives. As our MapReduce model has two outputs, MR – Ctrl will assure at the same time the response time and the availability specified in the SLA, while minimizing resource utilization.

The complete schema of the control architecture is presented in Fig. 20. All the variables used in the figure are defined in Table 1. More details regarding the implementation of the control framework can be found in [10]. As in Fig. 19, we consider the MapReduce system having two inputs (concatenated in the two dimensional vector *u*), one exogenous uncontrollable disturbance input *C* and two outputs (concatenated in the two dimensional vector *y*). Vector *u* contains the number of nodes in the cluster *N* and the max number of clients *MC*. While the y vector contains the response time *Rt* and availability *Av*.

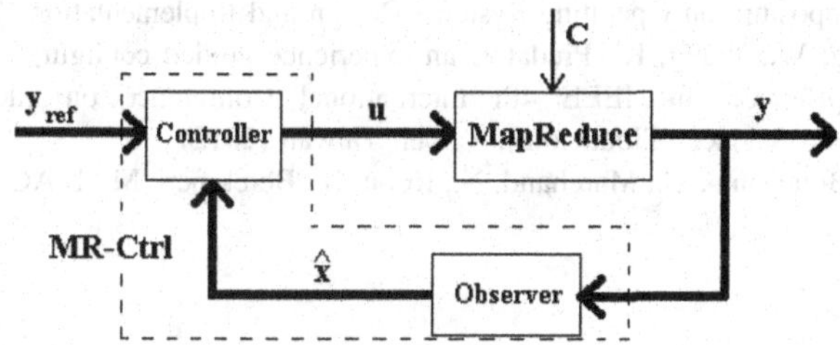

Fig. 20. The control architecture

Table 1. Definition of control variables

$y_{ref} = \begin{pmatrix} Rt_{ref} \\ Av_{ref} \end{pmatrix}$	Reference – response time and availability set in the SLA
$y = \begin{pmatrix} Rt \\ Av \end{pmatrix}$	Measured system output – response time and availability
$u = \begin{pmatrix} N \\ MC \end{pmatrix}$	System control input – number of nodes in the system and the maximum number of clients
C	Disturbance – number of clients trying to connect to the system
$\hat{x}$	Reconstructed behavior of MapReduce

5 Conclusions

This chapter describes the overall approach for managing SoS dynamicity. Its main intent is to associate a Service Level Agreement with SoS, and to provide SLA guarantees in terms of dependability, security, performance, etc. The overall MAPE Monitoring/Analysis/Planning/Execution approach is followed for SoS dynamicity management. The approach is illustrated through different implementations and techniques, e.g., a scalable monitoring, feedback control-based behavioural modelling and reaction strategies.

References

1. AMADEOS, Deliverable D3.1 - Overall Architectural Framework (2015)
2. AMADEOS, Deliverable D4.1 - Case study and use cases (2015)
3. AMADEOS, "Deliverable D4.2 - Case study realization," (2016)
4. Kappen, B.: Stochastic optimal control theory, Lecture Notes, Radboud University Nijmegen (2012)
5. Li, P.Y.: Advanced Control System Design, Ch. 6, Lecture Notes, University of Minnesota (2012)
6. Dreyfus, S.E.: Artificial neural networks, back propagation, and the Kelley-Bryson gradient procedure. J. Guidance Control Dyn. **13**(5), 926–928 (1990)
7. Theodoridis, S.: Machine Learning: A Bayesian Optimisation Perspective. Elsevier, London (2015)
8. Dean, J., Ghemawat, S.: MapReduce: simplified data processing on large clusters. In: USENIX Symposium on Operating Systems Design and Implementation (OSDI) (2004)
9. Lin, X., Tang, W., Wang, K.: Predator: an experience guided configuration optimizer for Hadoop MapReduce. In: IEEE 4th International Conference on Cloud Computing Technology and Science (CloudCom), Taipei, Taiwan (2012)
10. Serrano, D., Bouchenak, S., Marchand, N., Robu, B., Berekmeri, M.: IFAC World Congress (2014)

Case Study Definition and Implementation

Arun Babu[1], Francesco Brancati[1(✉)], Sorin Iacob[2], David Mobach[2], Marco Mori[3], and Thomas Quillinan[2]

[1] Resiltech SRL, Pisa, Italy
{arun.babu, francesco.brancati}@resiltech.com
[2] Thales Netherlands B.V., Hengelo, Netherlands
{sorin.iacob, david.mobach, Thomas.quillinan}@nl.thalesgroup.com
[3] Department of Mathematics and Informatics, University of Florence, Florence, Italy
marco.mori@unifi.it

1 Introduction

In this chapter we present three case studies in the smart grid domain: Electrical Vehicle charging, Household Management, and an integrated case study that combines the first two together with ancillary services. These case studies are first modelled using the AMADEOS Architectural Framework (AF) and associated tooling. We utilise the four levels of the AMADEOS AF: *mission*, *conceptual*, *logical* and *implementation*, as well as the seven viewpoints that have been defined: *Structure*, *Dynamicity*, *Evolution*, *Dependability and security*, *Time*, *Multi-criticality* and *Emergence*. We therefore examine the entire lifecycle of the framework considering some real-world case studies. These case studies are based on experts' feedback, including AMADEOS Advisory Board members, to ensure that realistic architectures are designed.

The architectures developed in this chapter will be further instantiated in a simulation environment, using the simulation tooling developed in the AMADEOS project. With these instances several experiments will be run in order to validate the framework as well as the architectures that were defined.

The three Smart Grid-based case studies described in this chapter are used to prove the effectiveness and consistency of the AMADEOS architectural framework. The method provided by the framework allows to design and implement a generic SoS in a procedural and systematic way. To accomplish this, the AMADEOS project defines a pyramidal top-down approach that must be undertaken passing through four different levels: a mission for the SoS, the conceptual level, where the ideas and concepts of the SoS are defined in order to support the capabilities of the SoS. Next, the logical level where the SoS is designed and these concepts are adapted towards supporting the requirements of the individual SoS domain. Finally, these are actualised in the implementation level, where the design is contextualized and realized in the enterprise.

This work has been partially supported by the FP7-610535-AMADEOS project.

A. Bondavalli et al. (Eds.): Cyber-Physical Systems of Systems, LNCS 10099, pp. 207–238, 2016.
DOI: 10.1007/978-3-319-47590-5_8

2 Smart Grid SoS

2.1 Smart Grid SoS Model

Figure 1 shows the model of the Smart Grid SoS created using the supporting facility tool. The Smart Grid SoS is composed of: EV_Charging, Medium_Voltage_Control, and the Household CS.

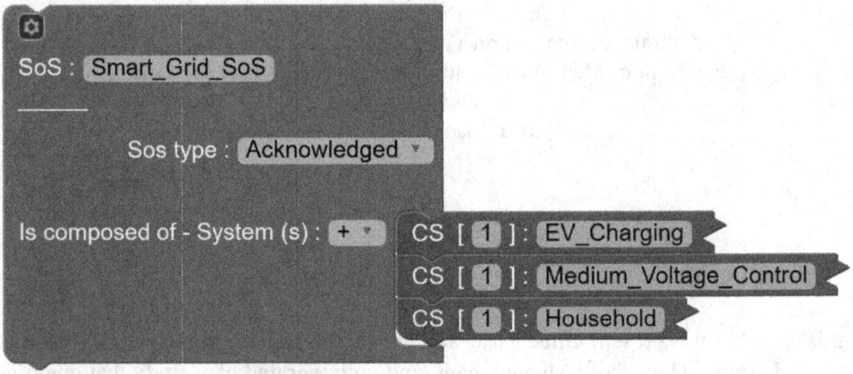

Fig. 1. The Blockly Smart Grid SoS model

2.2 EV-Charging SoS Case Study

Mission. The EV SoS must be designed to provide a friendly and convenient service to the users and at the same time, profitable to the provider. Planning and scheduling is of paramount importance for both energy providers and users: as an example, on one side, if the charging requests are spread during the day, there will be limited and/or controlled load peaks on the grid to be handled, thus the energy price may not vary abruptly over time and prioritized consumers (e.g., police and fire-fighter vehicles, ambulances, etc.) will be easily handled by the charging station operators. On the other hand, knowing the energy prices and available time slots, the users will be able to carefully plan the recharging operation while keeping the service affordable.

A typical scenario would be as follows: EVs travel through a wide area, where several charging station operators provide recharging services, by means of charging points. Drivers in need of power for their EV can provide the expected charging context (duration, power, etc.) to the e-mobility service in order to receive information regarding recharging time slots and associated energy prices of each charging station operator. A load management optimizer that cooperates with the charging station operators carries out planning and scheduling activities. The interested driver will then choose one of the slot-price pair possibility for recharging its vehicle, will be allowed to plug-in its EV at the charging point of the chosen charging station operator during the reserved time slot only, and the amount due will be based on the energy consumption times the booked price. At the end of the recharging operation, the driver receives a billing invoice.

It is evident from the above-mentioned scenario that various dependability aspects need to be considered. The participant systems need to be time synchronized to provide consistent information for scheduling and planning purposes. Furthermore, critical EVs should be prioritized for recharging with respect to any other vehicle. Therefore, each EV should be assigned a priority level. Moreover, the e-mobility service should be accessible by registered users only, i.e. the owner of an EV, to reduce the possibility of denial of service attacks being performed by illicit malevolent users scheduling recharging reservations, without a real need.

Conceptual Level. In this section we report the result of the activities performed at the conceptual level for the EV charging case study. For each viewpoint we list the most representative identified SoS requirements defined taking as input the SoS meta-requirements [1]. Traceability of the full set of requirements on meta-requirements is also provided in [4].

Architecture Viewpoint. The EVC consists of a subset of the CSs of the SoS "*Electrical Vehicle Charging in Smart Grids*" described in Chapter 2.1 of [4]. In particular (Table 1):

High level representation of EV-Component interactions.

A pictorial view of the EV charging SoS is reported in Fig. 2.

The steps required to recharge an EV are described below. Each step number corresponds to a sequence of actions that are carried out.

Table 1. EV charging case study components description

Name	Description
EV	An electrical vehicle.
Driver	The driver of the EV.
Charging Point (CP)	A physical connection for recharging the EV.
Smart Meter (SM)	A smart meter providing production and consumption values. May also enable advanced sensor facilities providing active/reactive production, frequency monitoring, voltage monitoring, etc.
Charging Station Operator (CSO)	Operator of a charging station, which is an electrified parking lot with several CPs, represented by an independent enterprise or owned by energy provider.
Load Management Optimizer (LMO)	The main software component connected to the grid which is in charge of providing power constraints and energy set points to the CSO.
E-mobility service	- Lists the best charging station locations to EVs, - Handles reservations. - Receives availability updates from CSOs.

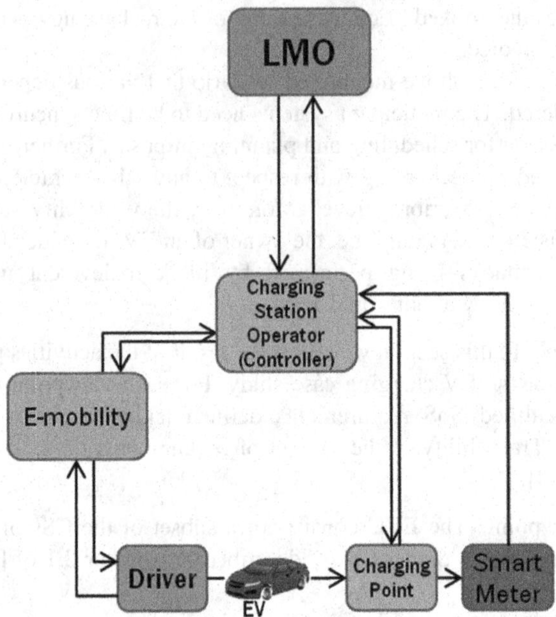

Fig. 2. Electrical vehicle charging SoS architecture

1. The information flow starts when a driver needs to recharge their EV. The driver requests a charging opportunity from the E-mobility service by providing the expected charging context (duration, power, etc.);
2. The E-mobility replies with the charging opportunity availabilities by accounting all CSOs present in the SoS;
3. Once the charging opportunities are received, the Driver asks for a reservation towards the most comfortable CSO according to his needs (e.g. availability of energy, distance, etc.) and the E-mobility forwards the message to the correspondent CSO;
4. The CSO updates its schedule, considering the received request and the power constraints defined by the set points provided by the LMO. The CSO reserves the desired time slot, allocates the resources and sends to the E-mobility service an acknowledgement;
5. The E-mobility service forwards the acknowledgment to the driver;
6. The driver reaches the CSO within the booked time slot. The EV is plugged in to the Charging Point (CP) and the CSO is notified about this event;
7. The CSO decides whether to allow the charging operation or to deny it proposing a re-scheduling of the time slot. This can happen if the situation has changed between the time the driver has booked their slot and they arrive at the CP. For example, if a higher priority EV has requested a slot;
8. While recharging, the Smart Meter measures the energy consumption and the resultant EV load for, respectively, billing and smart grid power flexibility purposes. Such information are sent to the CSO;

9. At the end of the charging cycle the EV is unplugged,
10. The CSO sends the billing invoice to the driver through the E-mobility service.

Dependability and Security Viewpoint. It is of paramount importance to guarantee the highest achievable availability and reliability of the SoS so that the charging requests are readily and continuously available. Achieving these properties allow not only to reach the highest quality of service, but also to maximise the profits. The SoS allows drivers to enquire for a charging opportunity through the E-mobility service and to have access to the CPs at any time. Thus, energy grid enterprises can maintain the grid stability, balance the load, and ensure correct voltage levels and frequency, etc., while energy providers can make the highest profits out of it. To avoid undermining such premises, Denial of Service (DoS) attacks caused by unauthorized users enquiring the E-mobility service or trying to access CPs need to be properly tackled.

Dynamicity Viewpoint. EVs join and leave the SoS, according to the need for charging of the EV. The change in the topology due to this turnover of EVs characterizes the dynamicity (DYN) of the SoS.

Emergence Viewpoint. According to [2], Emergence is: *A phenomenon of a whole at the macro-level is emergent if and only if it is new with respect to the non-relational phenomena of any of its proper parts at the micro level.* This indicates that these phenomena cannot be observed at CS level, but at SoS level (or other higher level). As consequence of identification of emergence scenarios, hazard analysis must be performed to identify and mitigate any hazards.

Evolution Viewpoint. Due to technological advances, marketing, or customer needs; different kind of EVs it may be necessary to improve, change or add services.

Multicriticality Viewpoint. The scenario described so far, only foresees normal EVs (e.g., private EVs) that need to recharge and ask for charging opportunity. However, during real-life emergencies, e.g. rescue operations, wildfires and public security, the requirement is that specific EVs must always be available. Therefore, the CSO scheduling strategies must prioritize such vehicles before any others. Emergency EV drivers access prioritized charging in the same fashion of any other driver, i.e. through the E-mobility service.

Time Viewpoint. As a general remark, it is worth noting that most of the information exchanges between CSs rely upon a notion of time. As an example, it would be impossible to plan and schedule a request of recharging an EV by a driver for the current day or even pay the billing invoice if not properly time-stamped. Thus, to provide the payment services, to allow users to enquire for charging opportunities and to effectively, and efficiently, plan, and schedule, recharging operations over time, there must be awareness of time over the SoS. Further, each CS must be time synchronized to a common reference time to successfully provide their services.

Logical Level. This section describes the SoS Logical Description of the SoS defined in [4] using the model made using the supporting facility tool.

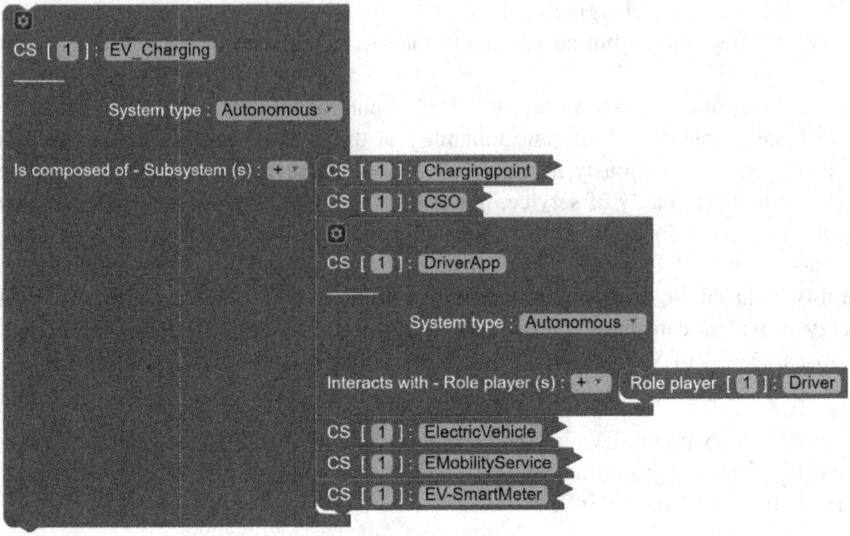

Fig. 3. EV Charging in Blockly

After loading the model in supporting facility and double clicking on EV_Charging CS we can see in Fig. 3 that the EV-Charging Blockly model consist of CSs that matches to the diagram depicted in Fig. 2.

Figures 3 and 4 show that EV-Charging CS consists of the following CSs: (i) Chargingpoint, (ii) CSO, (iv) DriverApp, (v) ElectricVehicle, (vi) EMobilityService, and (vii) EV-SmartMeter

In the following section will be described in details the CSs listed above, through the expansion of the blocks.

Charging Point CS. The Chargingpoint CS can be expanded by double clicking on it and the result is depicted in Fig. 4.

Figure 4 shows that the Chargingpoint CS communicates with the Electric Vehicle, CSO, EV-SmartMeter CSs by the ChargінPoint-ElectricVehicle RUPI, Chargingpoint-CSO RUPI and Chargingpoint-EV-SmartMeter RUPI, respectively. Figure 4 also shows that the services provided by the Chargingpoint CS. Furthermore, it has a State Variable Chargingpoint: charging_done.

From the viewpoint of communication, double clicking on RUPIs and RUMI blocks of Charging Point, CS Fig. 4 shows that the Chargingpoint CS provides the services to CSO CS through the Chargingpoint-CSO RUMI. The Fig. 4 shows also that:

- The Chargingpoint-ElectricVehicle RUPI transports the Plug-OUT-Signal and it is connected to the ElectricVehicle-ChargingPoint RUPI of the ElectricVehicle CS,
- The Charging Point-EV-Smart Meter RUPI transports Electricity and it is monitored through the Charging Point Probe.

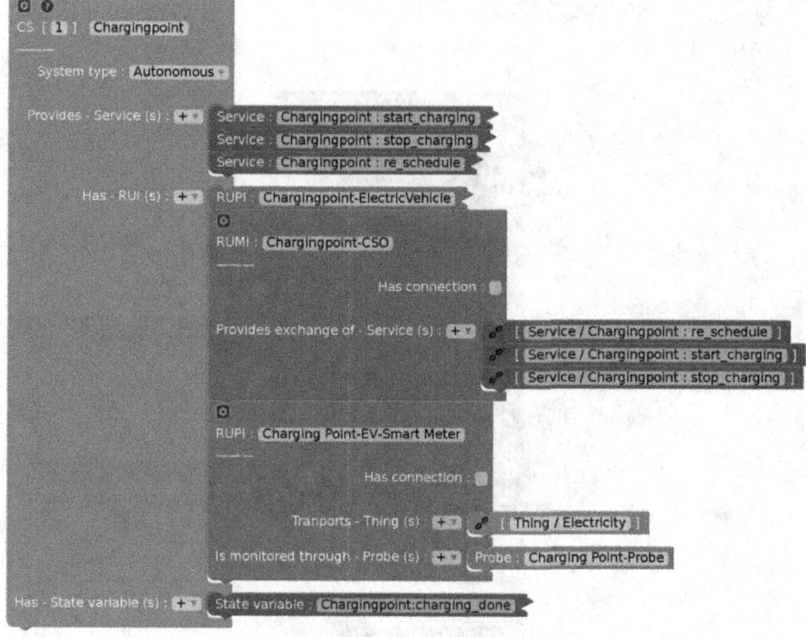

Fig. 4. The Blockly Charging point CS model

CSO CS. On expanding the CSO CS block (Fig. 5), its services, RUIs, MAPE, and state variables can be seen.

Figure 5 shows that the Chargingpoint CS communicates with the EmobilityService, Chargingpoint, LMO, Aggregator CSs by the CSO-EMobilityService, CSO-Chargingpoint, CSO-LMO, CSO-Aggregator RUMIs respectively. The Fig. 5 also shows the services provided by the CSO CS. It also has a State variable CSO: reservation.

From the viewpoint of communication, double clicking on RUMIs blocks of CSO CS, Fig. 5 shows that the CSO CS provides the following services:

- CSO:do_charging_reservation, CSO:do_priority_charging_reservation to the EmobilityService CS through the CSO-EMobilityService RUMI,
- CSO: EV Charging Schedule and CSO: Update energy consumption to the LMO CS through the CSO-LMO RUMI
- CSO: Set Energy price and CSO: Forward energy price to the Aggregator CS through the CSO-Aggregator RUMI

Figure 5 also shows that the CSO-Chargingpoint RUMI is connected to the Chargingpoint-CSO RUMI in order to call the service provided by ChargingPoint CS; the CSO-LMO RUMI is monitored through the Probe CSO Probe. The CSO has a local clock that we call CSO-clock and that will be described when addressing the Time viewpoint.

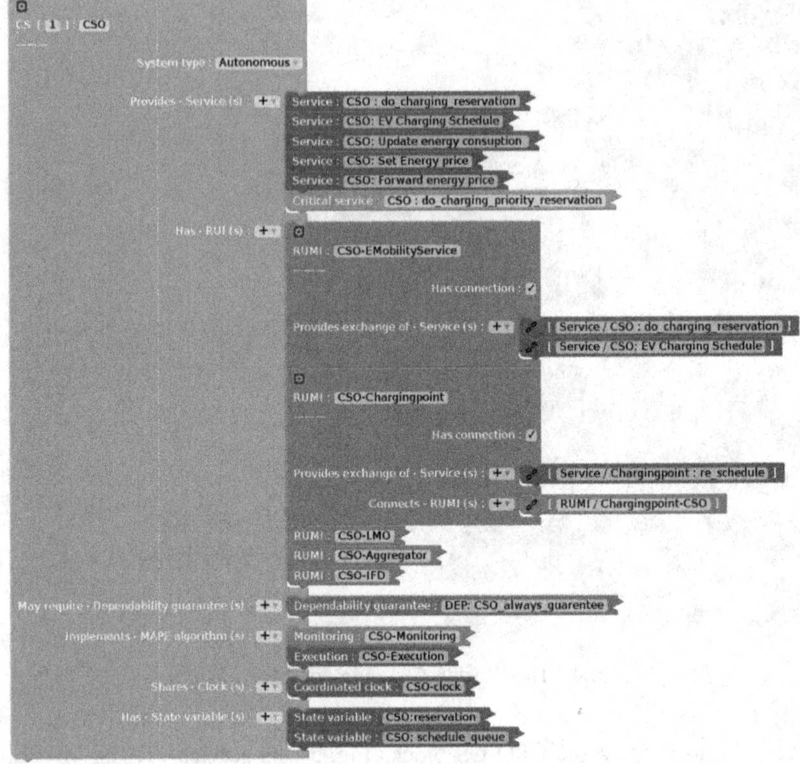

Fig. 5. The CSO CS Blockly model

DriverApp CS. On expanding the DriverAPP CS block in Fig. 6, it shows that the DriverApp CS communicates with EMobilityService CS by the DriverApp-EMobilityService RUMI, and shows that provides the service DriverApp: accept_reservation.

This figure also shows that DriverApp CS:

- Interacts with a Role player: Driver,
- Has the following State variables:
  - DriverApp:result_of_charging_opportunities_request;
  - DriverApp: selected_charging_opportunity;
  - DriverApp:duration;
  - DriverApp:power,
  - DriverApp:got_reservation.

From the viewpoint of communication, double clicking on DriverApp-EMobilityService RUMI, we see that the DriverApp CS provides the service DriverApp:accept_reservation by DriverApp-EMobilityService RUMI to EMobility CS, and

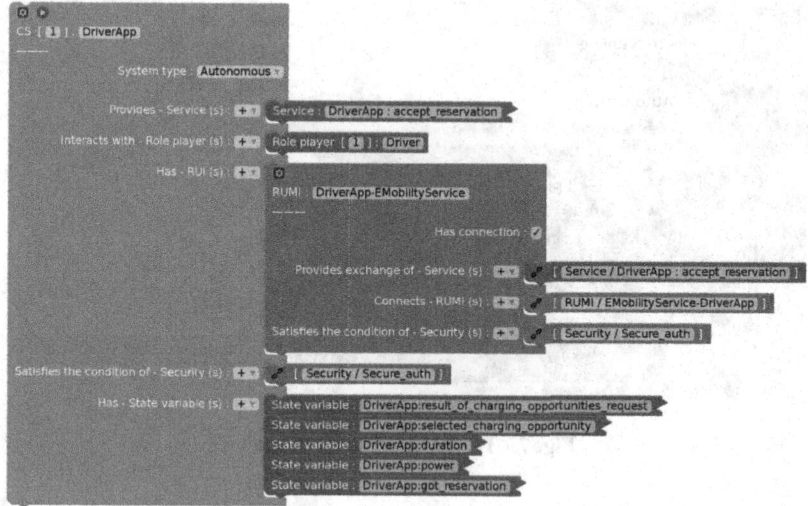

Fig. 6. The DriverAPP CS Blockly model

this RUMI is connected to the EMobilityService-DriverApp RUMI in order to call the services provides by the EMobilityService.

ElectricVehicle CS. On expanding the ElectricVehicle block as depicted in Fig. 7, it is shown that the ElectricVehicle CS has ElectricVehicle-Chargingpoint RUPI in order to connect with the Charging Point CS. Indeed, expanding the ElectricVehicle-Chargingpoint RUPI you can see that the RUPI transport Plug-In-Signal and it is connected with Chargingpoint-ElectricVehicle RUPI.

EMobilityService CS. On expanding the EMobilityService CS block as depicted in Fig. 8, it is shown that the EMobilityService CS communicates with the DriverApp, Aggregator, Market, CSO CSs by the EMobilityService-DriverApp, EMobilityService-Aggregator, EMobilityService-Market, EMobilityService-CSO RUMIs respectively.

Figure 8 shows also that EMobilityService CS provides the services:

- EMobilityService: Set Energy price
- EMobilityService: Forward Energy price
- EMobilityService: do_charging_reservation
- EMobilityService: get_available_charging_opportunities
- EMobilityService: re_schedule

and has the State variables:

- EMobilityService:available_charging_opportunities
- EMobilityService:charging_op_sent_by_driver
- EMobilityService:reservation_to_be_sent_to_driver

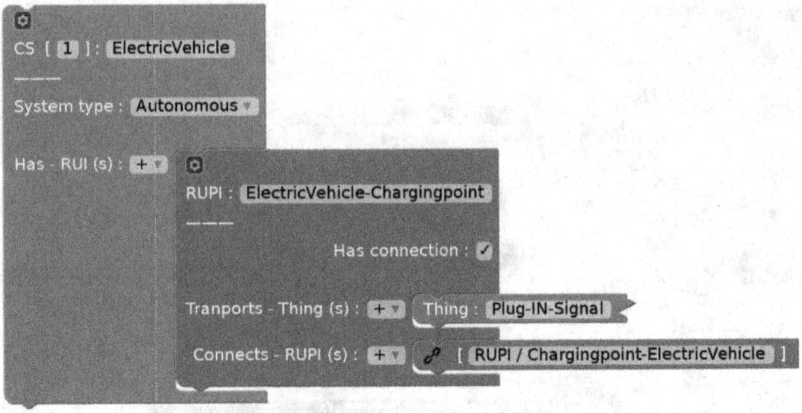

Fig. 7. The EV CS Blockly model

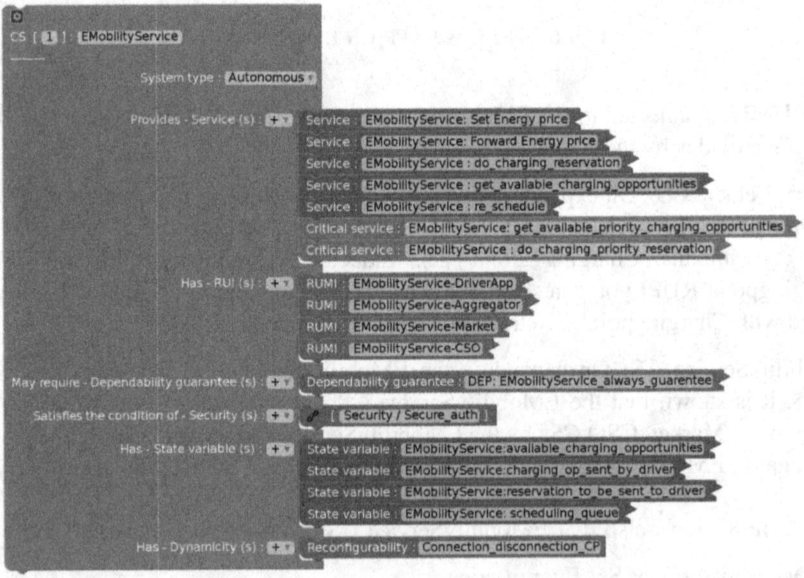

Fig. 8. The EMobility CS Blockly model

From the viewpoints of communication, double clicking on RUMIs blocks of EMobilityServices CS, the Fig. 9 shows that the EMobilityService CS provides the services:

- EMobilityService:do_charging_reservation, EMobilityService: get_available_charging_opportunities, EMobilityService: re_schedule through the EMobility Service-DriverApp RUMI

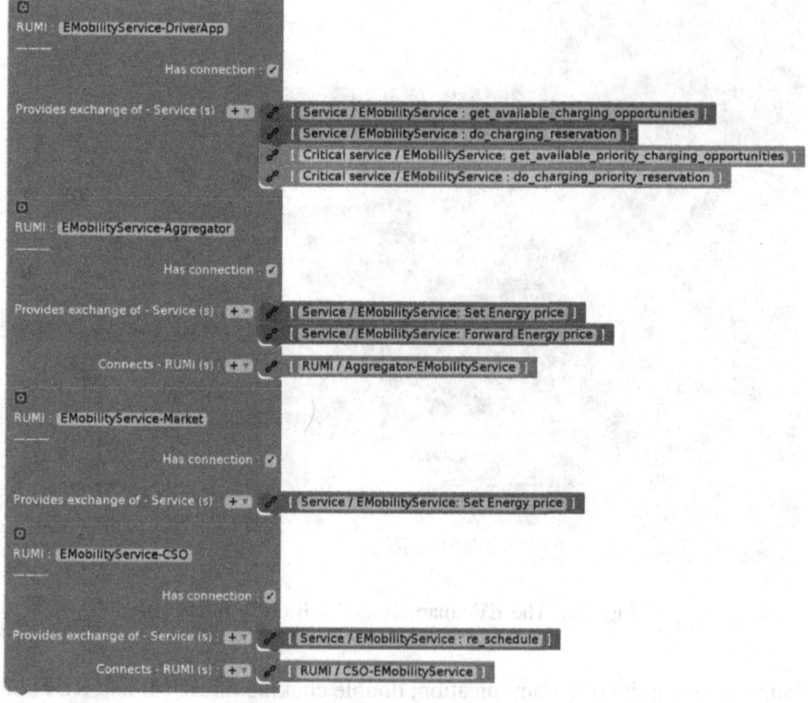

Fig. 9. The EMobility CS RUMI model

- EMobilityService: Set Energy price, EMobilityService: Forward Energy price through the EMobilityService-Aggregator, that is connected to the Aggregator-EmobilityService RUMI
- EMobilityService: Set Energy price through the RUMI EMobilityService-Market.

Furthermore, the EMobilityService-CSO is connected to the CSO-EmobilityService RUMI in order to call the services provided by CSO CS. The Fig. 9 shows also that EMobilityService has the State Variable:

- EMobilityService:available_charging_opportunities
- EMobilityService:charging_op_sent_by_driver
- EMobilityService:reservation_to_be_sent_to_driver.

EV-Smart Meter CS. On expanding the EV-SmartMeter CS block as depicted in Fig. 10, it shown that the EV-SmartMeter CS communicates with Chargingpoint and Meter Aggregator CSs by the EV-Smart Meter-Chargingpoint RUPI and EV-Smart Meter-Meter Aggregator RUMI respectively.

Figure 10 shows also that the EV-SmartMeter CS provides the service EV-Smart Meter: Get_energy_consumption and implements MAPE Algorithm.

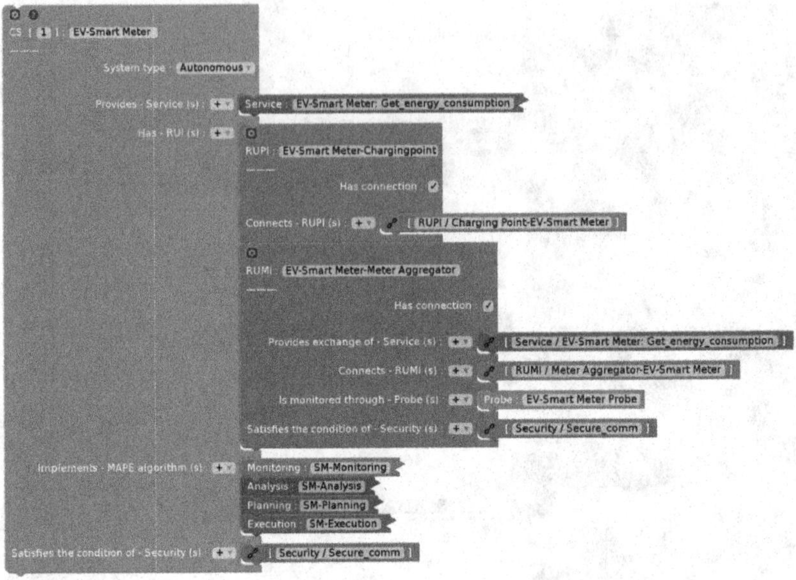

Fig. 10. The EV-smart meter CS Blockly model

From the viewpoint of communication, double clicking on RUMI and RUPI blocks of EV-Smart Meter CS, Fig. 10 shows that the EV-SmartMeter CS provides the service EV-Smart Meter: Get_energy_consumption through the EV-Smart Meter-Meter AggregatorRUMI, and the EV-Smart Meter-Chargingpoint RUPI is connected to the Charging Point-EV-SmartMeter RUPI of the Chargingpint CS.

Implementation. As stated in D4.2 [4], the EV SoS must be designed to provide a friendly and convenient service to the users and, at the same time, profitable to the provider. Planning and scheduling is of paramount importance for both energy providers and users: as an example, on one side, if the charging requests are spread during the day, there will be limited and/or controlled load peaks on the grid to be handled, thus the energy price may not vary abruptly over time and prioritized consumers (e.g., police and fire-fighter vehicles, ambulances, etc.) will be easily handled by the charging station operators. On the other hand, knowing the energy prices and available time slots, the users will be able to carefully plan the recharging operation while keeping the service affordable.

A typical scenario would be as follows: EVs travel through a wide area, where several charging station operators provide recharging services, by means of charging points. Drivers in need of power for their EV can provide the expected charging context (duration, power, etc.) to the e-mobility service in order to receive information regarding recharging time slots and associated energy prices of each charging station operator. A load management optimizer that cooperates with the charging station operators carries out planning and scheduling activities. The interested driver will then choose one of the slot-price pair possibility for recharging its vehicle, will be allowed to plug-in its EV at the charging point of the chosen charging station operator during the reserved time slot only, and the amount due will be based on the energy consumption times the booked price. At the end of the recharging operation, the driver receives a billing invoice.

With this scenario in mind, and using the AMADEOS tooling, we designed a SoS (shown in Fig. 11, and described in D4.1 [3]) based on a typical EV rollout, in particular based on the desired situation in the Netherlands, based on interviews and workshops with experts, both in the EAB and Grid operators. This SoS was modelled using the Blockly tool (this is described in D4.2 [4]) and a simulator was generated from the tool. We combined this simulator with a simulation toolkit called SimPy and performed a number of experiments based on the scenarios defined in D4.2. This served to both validate the simulator as well as determine if such a simulation could be used to determine and validate possible future designs of the EV charging network, based on varying user and SoS behaviour.

The simulation is built using "SimPy" and uses fixed time slots for reservations. At the moment these slots are 15 min long and start on 0, 15, 30 and 45 min past each hour. Each Driver that wants to charge will attempt to make a Reservation for a Charging Point (via the E-Mobility Service) at a CSO. Because of these timeslots, a Reservation will only begin at the start of the next timeslot and will last for a number of time slots as calculated by the CSO. After the reserved time has passed, the Driver is expected to Plug Out its EV from the Charging Point, though he can do so at an earlier point in time (if the EV is already fully charged, or the driver otherwise decides to do so).

Below, we summarize some of the results received. These are preliminary results, and as the data retrieved from the simulator is extremely extensive, we identify only some of the more interesting aspects. In general, we found the tooling to be extremely useful, and, for example, several emergent behaviours were discovered in the data that, if proven true, would lead to significant challenges to the electrical grid. Furthermore, the aspects that the SoS was designed to test (primarily security and dependability) were successfully tested and validated. We plan to exploit these results in two manners: First to follow up on the results and perform more experiments after the end of the project, leading to publications, and secondly both Thales and ENCS are planning to make the simulator code available online, so other researchers can validate and use the same simulation code.

Results from the EV Charging Scenarios. The first scenario that we will discuss is a usage case where EV drivers do not immediately remove their vehicles from the charging points for a period of time. This means that those drivers who do not remove their EVs are acting badly – blocking charge points from other users. We chose a fixed period of 4 h for this behaviour in this scenario, and the simulation took place over a period of 24 h (simulated time via SimPy). There were 1000 EVs present and 500 Charging Points (CPs). In this simulation, market costs did not cause a significant change in behaviour. The EVs were set to desire a charge of 70%, with a "must charge" threshold at 20%. Four states are defined in the simulation: Charging, Driving, Idling and Waiting. Charging and Driving are self-evident. Idling was periods of time when the EV was not in use (due to lack of driver need). Waiting was an undesirable state where an EV was waiting for a CP to become available. In order to stress the Grid and

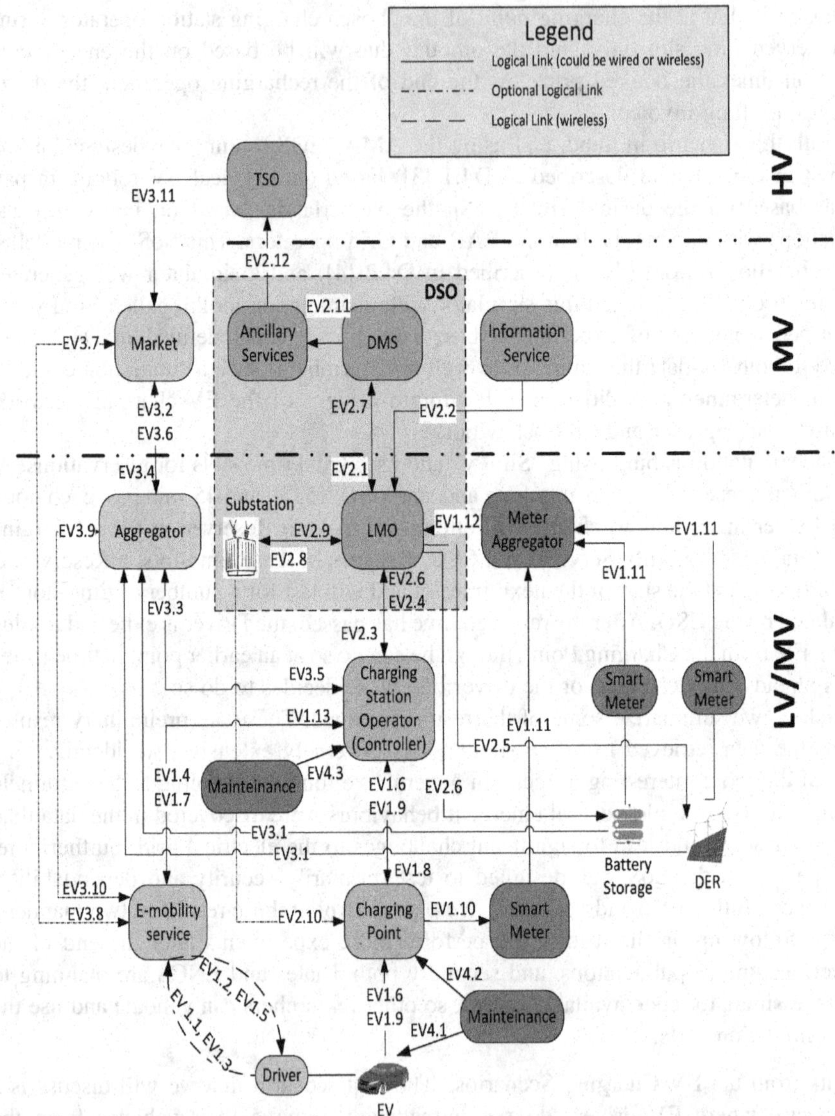

Fig. 11. EV Charging SoS

the SoS as a whole, Idling was the least desired state of the three normal states – the chance of an EV idling was set to 10%. Finally, note that the drivers, cars and initial state were randomly assigned and a period of 15 min was allowed to let the state settle. This can be seen in an initial jump in all of the graphs.

In Fig. 12, you see four basic types of behaviour: Driving, Idling, Waiting (for a free Charging Point) and Charging. The numbers on the X-axis reflect the number of EVs in that state. In this scenario, (and in all our tests, due to global variables that were

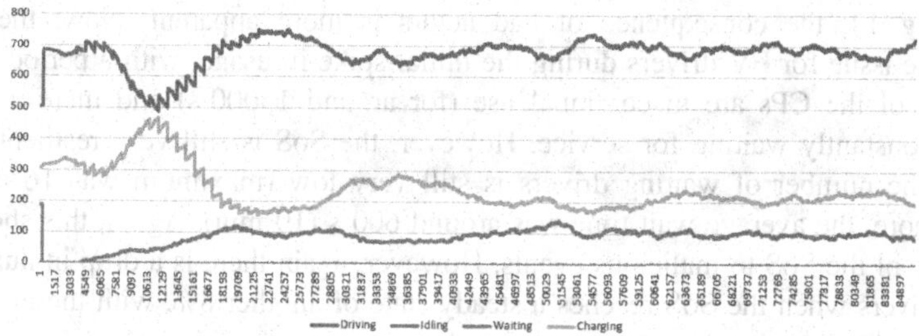

Fig. 12. Initial state where no drivers act badly

set) the initial state was close to optimum for each of the EVs, with around 70% driving and 30% charging. The initial spike in charging (around 10000 s into the simulation) is due to the initially driving EVs falling below the 20% lower threshold and requiring a charge. You can then see that after around 20000 s, the system reaches a steady state with around 70% driving, 20% charging and 10% idle. Note also that there is an insignificant number of EVs in the waiting state.

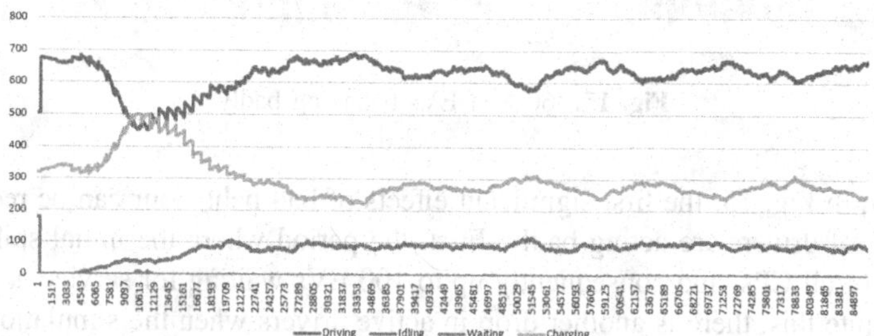

Fig. 13. 20% of EVs acting badly

Figure 13 shows the first set of EVs acting badly – in this case, 20% delay their disconnection from the CPs and block other drivers from using them. This scenario shows the resilience of the SoS to such behaviour – there is again the same spike in charging after around 10000 s, and a minor amount of EVs waiting, but again after around 20000 s, the SoS reaches a relatively steady state, although with more EVs charging (around 5% more) at any given time (and consequently 5% less driving) that the optimal case. One last thing to note is that all CPs are in use during the initial spike.

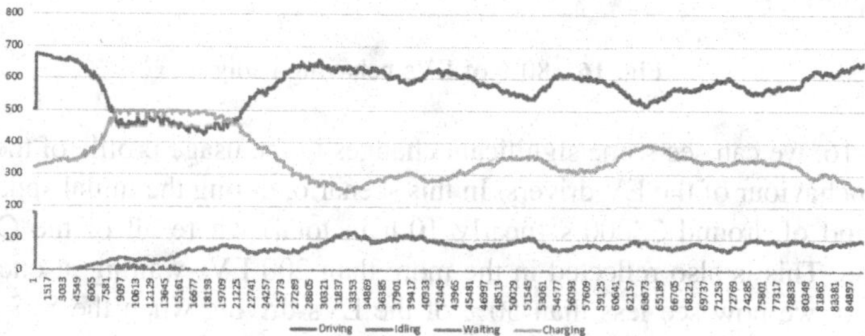

Fig. 14. 40% of EVs behaving badly

In Fig. 14, the consequence of bad actors is more apparent. Now, there is a noticeable issue for EV drivers during the initial spike in usage, with a period of time when all of the CPs are in continual use (for around 15000 s) and there are some drivers constantly waiting for service. However, the SoS is still very resilient to this issue – the number of waiting drivers is still very low (maximum was 16 drivers). Furthermore, the average wait time was around 600 s (10 min). Again, this shows the resilience of the SoS to malicious events. However, again there is a drop in number of active drivers when the SoS reaches a steady state of another 8%, with the number of charging drivers up by the same amount.

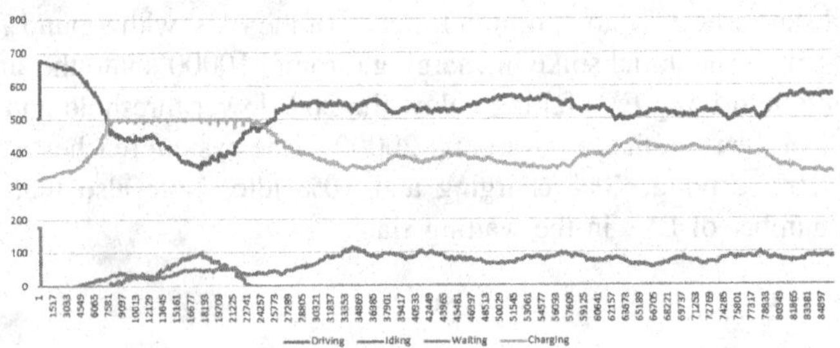

Fig. 15. 60% of EVs behaving badly

Finally, in Fig. 15, the first significant effects of bad behaviour can be recognized, while 60% of drivers are acting badly. First, the period where the initial spike causes full usage of all CPs, and consequently up to 100 EVs waiting for service at the worst point. Despite this, there is another drop in active drivers when the simulation reaches the steady state – down to around 58% from a high of over 70% in the control simulation.

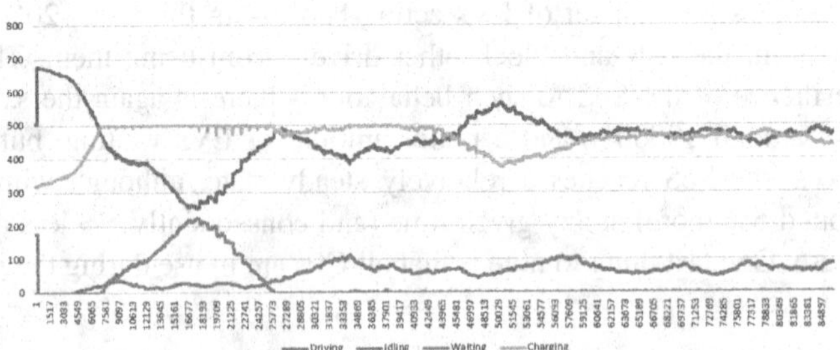

Fig. 16. 80% of EVs behaving badly

In Fig. 16, we can see some significant changes in the usage profile of the SoS due to the bad behaviour of the EV drivers. In this scenario, during the initial spike, there is now a period of around 35000 s (nearly 10 h in total) where all of the CPs are in constant use. This is also reflected in the more than 200 EVs waiting at one point. In this scenario, we now see less than 50% of the EVs driving when the SoS reaches a

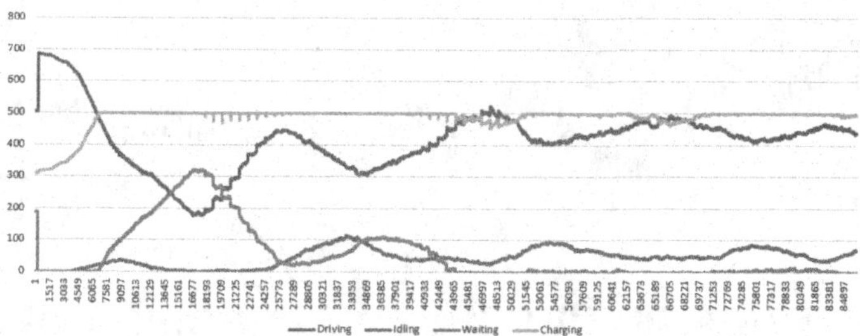

Fig. 17. 100% of EVs behaving badly

steady state, with roughly the same number charging as driving. This means that 20% of the EVs have changed from driving to charging since the control simulation. However, this is taking place where only 20% of drivers are behaving correctly and yet the SoS remains (for the most part) available for use.

In the final simulation, shown in Fig. 17, every driver is now acting badly. In this case, the effect on the SoS is dramatic and relatively catastrophic. In this simulation, for essentially the entire day, all of the CPs are in constant use, the number of drivers is down to less than 50% and for large periods of time, EVs are waiting to charge (more than 30% at the peak).

Market Simulation and Energy Usage. These simulations were intended to determine the reaction of the SoS to malicious behaviour on the part of the drivers. The results described above shows how the SoS behaves from the perspective of the EV drivers. However, there is another important aspect that can also be studied: the reaction of the SoS from the perspective of the Grid. This was calculated from the perspective of the TSO (see Fig. 11). The mission of the SoS is, basically, to ensure the stability of the Grid, and to ensure that large changes in energy generation are not required. In order to achieve stability, a number of measures were enacted. First, the CPOs (there are 5 independent CPOs in this simulation) received a wholesale price from the DSO, via the market, based on:

1. Forecasted demand (by the DSO) for the next 24 h, in 15 min intervals, and
2. How much energy they predicted that they required also in next 24 h, in 15-minute intervals.

The goal of a CSO was, using the market price, to ensure they did not stray too far from their forecasted need. Prices were set by the DSO based on five energy bands – the price per unit was set based on the band that the CSO requested, regardless of the actual energy used in reality. The ideal situation for a CSO was to get as close to the top of a band, without exceeding it, as the cost per unit would jump, and make the per unit cost more expensive. Therefore, if a CSO discovered that (based on EV reservations) that they were going to jump up to the next band, their behaviour would be to attempt to get many more customers, by reducing the customer price. We used this aspect to drive competition and the response from the domain experts was that this is indeed a desired future scenario. This aspect proves the evolutionary promise of an AMADEOS SoS design.

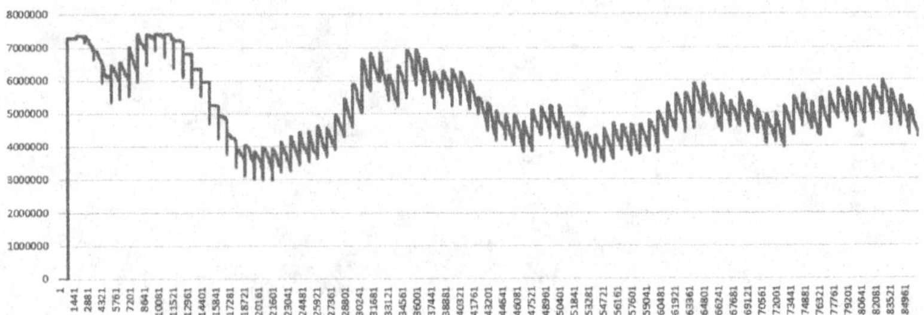

Fig. 18. Total network load with 0% bad behaviour.

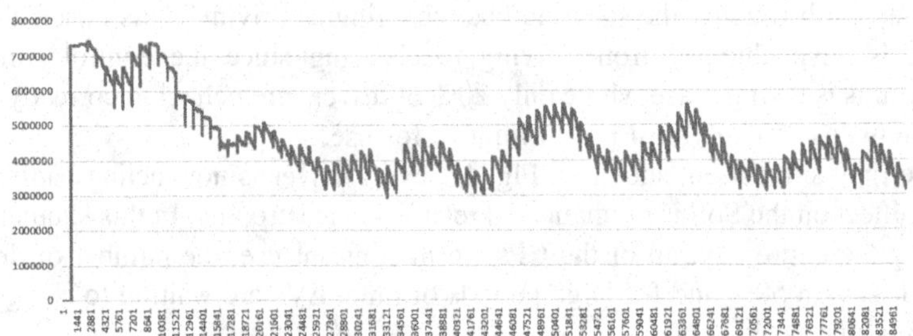

Fig. 19. Total network load with 40% bad behaviour

Based on the same simulations as shown in Section 0, Fig. 18 shows the reaction of the Grid to the EV charging scenario where no bad behaviour is present. The requests come in 15-minute intervals, and this can be seen in the jagged lines that are present in the graphs. In this instance, there is again an initial jump, after the 15 min settling down period and then a peak and trough after around 18000 s. This is again due to the jump in number of users charging at the start of the simulation based on their initial charge state and desire not to fall below 20%/attain 70% charge. The interesting outcome is the narrow band (between 40 MWh and 60 MWh) that the Grid eventually stabilizes towards. Future research will definitely consider running such experiments over several weeks of simulated time and integrating a typical Grid usage pattern into the demand[1].

The 20% (not shown) and 40% bad behaviour results (see Fig. 19) show how the bad behaviour is increasingly reflected in the variations of load over time. The intermediate results (60% and 80% bad behaviour) show increasingly wide variations, cumulating to the wide swings shown in Fig. 20, where all of the drivers are again behaving badly. These variations once again show how malicious actors will cause significant issues for Grid operators.

[1] As discussed in D4.2, the simulation was not run in concert with the MV case study.

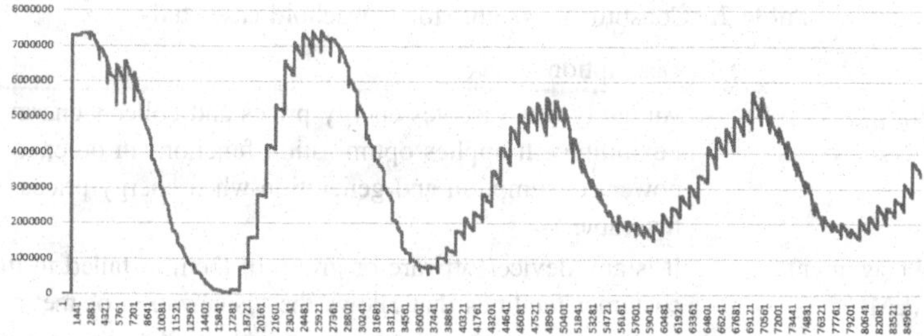

Fig. 20. Total network load where 100% of the drivers are acting badly

2.3 Household Management Case Study

Mission. The goals of the SoS are somewhat similar to the ones of the EVC: allow users, i.e. households, to use home appliances (e.g., flexible or general loads) at a convenient price while properly scheduling the energy consumption and distribution to improve revenues for energy providers with the constraint of keeping a balanced load on the grid.

During a normal scenario, a user wants to activate one or more home appliances, thus requiring for energy from the grid. Energy requirements are managed by an energy management gateway every time a person wants to activate home appliances. The energy management gateway handles the scheduling of energy consumption/provision within the household and sends requests to the coordinator to update consumption/production setpoints. The coordinator continuously updates the energy setpoints according to the grid energy availability. The actors need to be time synchronized to provide consistent information to the coordinator for scheduling and planning actions. Furthermore, some home appliances should be energy prioritized with respect to others (e.g., refrigerators, air conditioning systems, etc.).

Conceptual Level. In this section we report the result of the activities performed at the conceptual level for the Household management case study. For each viewpoint we list the most representative identified SoS requirements defined taking as input the SoS meta-requirements [1]. Though, the requirements come from meta-requirements, the traceability of the full set of requirements on meta-requirements is not provided. Some of the representative requirements are given below (the full set of requirements can be found [4]).

Viewpoint Examination. The objective for which the SoS is designed for is to allow end customers (i.e., households) to interact with the coordinator in order to request the activation of some particular appliance.

Architecture Viewpoint. The HHM consists of a subset of the CSs of the SoS "Household scenario" described in Chap. 2.2 of [4]. In particular (Table 2):

Table 2. Constituent systems for Household case study

Name	Description
Coordinator	An entity that receives energy prices and collects energy flexibilities. It applies optimization functions in order to shift power consumption and generation when energy prices are favorable.
Energy Management Gateway (EMG)	It is any device/software or group of them installed in the customer facilities that allows the visualization of metrological information, price and warning signals by the customer and has the capability to take action (e.g. rescheduling of power consumption/production) automatically or after approval by customer on any home appliances.
Smart Meter (SM)	A smart meter providing production and consumption values. May also enable advanced sensor facilities providing active/reactive production, frequency monitoring, voltage monitoring, etc.
Distributed Energy Resource (DER)	DER devices are generation and energy storage systems that are connected to a power distribution system.
Home Automation Device	Device providing additional functionalities enabling consumers to interact with their own environment (e.g., a smart thermostat).
Flexible Load	Load that can be controlled by the EMG (e.g., a smart washing machine).
Local Network Access Point (LNAP)	Provides the WAN connection for upload of the metering data.
Display	Main Human Machine Interface (HMI) between the householder and smart services.

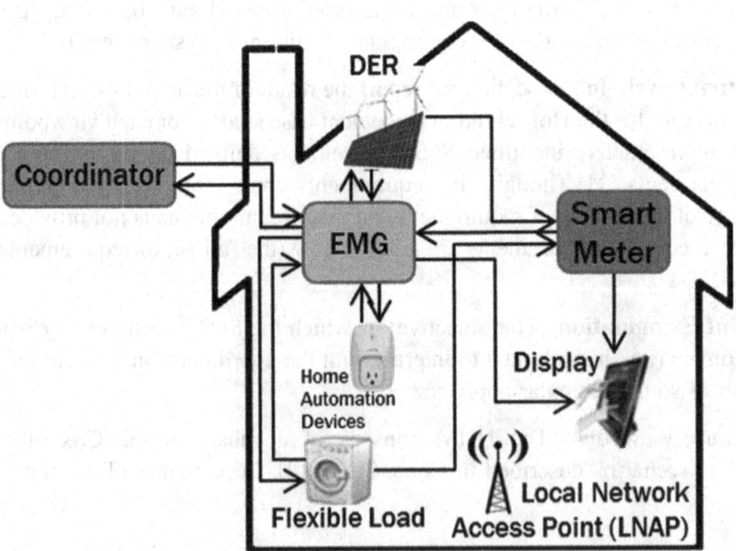

Fig. 21. Household management SoS architecture

High level representation of HH Management interactions.

A pictorial view of the HH management SoS is reported in Fig. 21.

Figure 21 shows a pictorial view of the temporal sequence 1–7 comprising, for each step, the involved systems and connections from Fig. 10.

In the following, the steps performed in a nominal household scenario are described. Each numbered step corresponds to a specific, adimensional, time instant at which the corresponding actions are carried out.

1. A person wants to activate the Flexible Load appliance. The Flexible Load sends an energy request to the EMG.
2. The EMG sends an aggregated energy request to coordinator.
3. The Coordinator decides whether to accept or reject the EMG request by updating energy setpoints according to the energy availability on the grid.
4. According to the Coordinator reply message, the EMG sends an OK/KO message to Flexible Loads Home Automation Devices and DERs.
5. In the case of an OK message, the Flexible Load is activated.
6. The SM measures the energy consumption and the resultant load for, respectively, billing operations and smart grid power flexibility purposes. Load information are forwarded to the EMG.
7. When the Flexible Load ends its tasks, the SM sends the total energy consumption and billing invoice to the user through the Display.

Dynamicity Viewpoint. The aforementioned case study is therefore very dynamic, in the sense that the service provided by the electrical grid is always changing, according to the customer requests for home appliances.

Evolution Viewpoint. Due to technological advances and new customer needs, the EMG should have the capability of easily integrate new HMIs, automation devices and loads. As an example, an householder may want the same interaction provided by the in-house Display on his smartphone and to use it not only while connected to the LNAP, but also when outside to schedule a washing machine while connected to the mobile carrier.

Multicriticality Viewpoint. Up to this point, we have described the case study assuming all the appliances with the same criticality. However, an HH consists of a large variety of possibly interconnected (net-centric) appliances providing different kind of services and, thus, characterized by different priority and criticality levels.

2.4 Medium Voltage Control SoS Case Study

Mission. The mission is to provide charging services to EV drivers and to provide energy-related services to households, but also to highlight interesting emergent phenomena that would not arises within the single SoS improving their interoperability with predictable, dependable behaviour avoiding negative cascading emergent effects.

Conceptual Level. The Medium Voltage energy distribution infrastructures are needed to interconnect the EV-Charging SoS and HH Management SoS and consist of the following set of CSs (Table 3):

Table 3. MV Energy distribution constituent systems

Name	Description
DER	Device able to produce energy in the grid.
Battery Storage	Battery used to store energy produced by DER
Smart Meter	A smart meter providing production and consumption values. May also enable advanced sensor facilities providing active/reactive production, frequency monitoring, voltage monitoring, etc.
Meter Aggregator	It is in charge of collecting metering data from the supervised smart meters.
LMO	The main software component connected to the grid which is in charge of providing power constraints and energy set points to the CSO.
DMS	A system which provides applications to monitor and control a distribution grid from a centralized location, typically the control centre. A DMS typically has interfaces to other systems, like a GIS (Geographical Information System) or an OMS (Outage Management System).
Information services	Commonly available services provided by a third party. E.g. weather information needed to predict PV production.
Substation	Substation system implementing the automation sequences and the control functions of interfacing process level control devices
Ancillary Services	Information services to TSO (e.g. extreme increment or decrement of the electricity frequency).
TSO	Operating centre for supervising critical regions of the transmission grid
Market	Set of services designed to evaluate the Energy price.
Aggregator	Is designed to provide information from the grid to the Market. The Aggregator is able to aggregate information incoming from sources reducing complexity and redundancy.

This section describes at logical level the infrastructures related to the Medium Voltage energy distribution (Fig. 22) and the examples of emergent phenomena that could be coming from the interoperability of Ev-Charging SoS and HH Management SoS.

In the following, the steps performed to manage the energy distribution are described.

1. Smart meters (SM) are devices connected to each prosumer;
2. SM collects and send information (also on demand) about power consumption/generation to the Meter Aggregator;
3. The Meter Aggregator sends the aggregated measures to the LMO (SM can provide information about energy consumption/generation to the EMG and the home Display);
4. The LMO receives information about Substation Monitoring Data, Household forecasted energy consumption generation and EV charging schedule;
5. The Distribution Management System (DMS) updates periodically LMO information for high level operation objectives, changes in data models (e.g. grid topology, newly connected charging station);

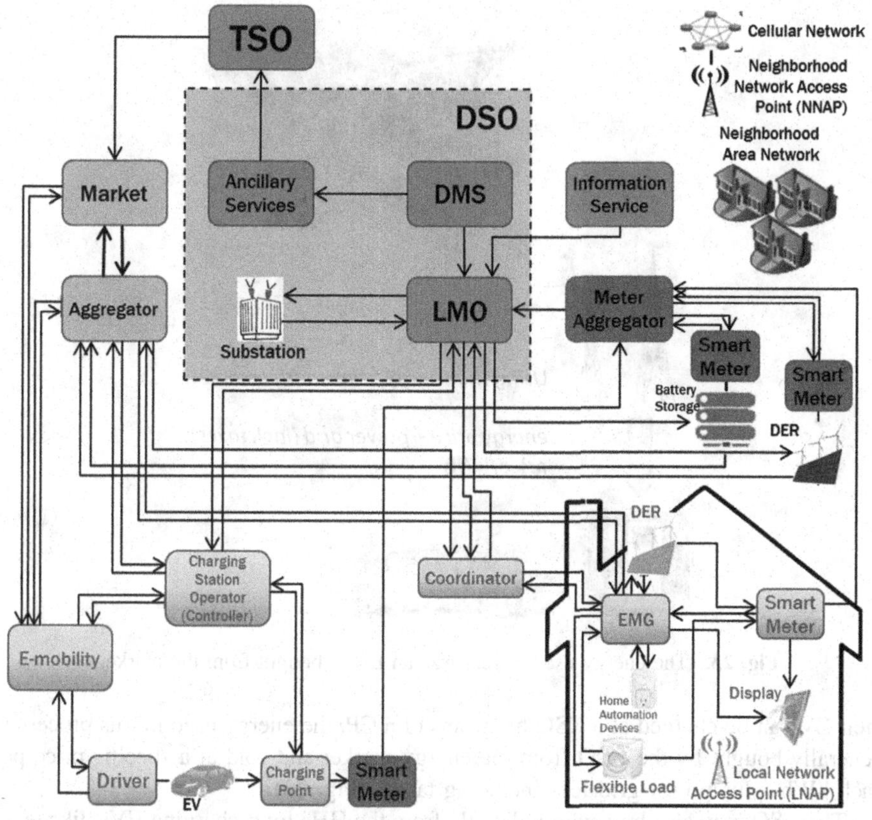

Fig. 22. Architectural view of Medium Voltage Control (3).

6. Information services updates LMO with environmental information (e.g. weather data)
7. The LMO sends periodically updates on energy consumption/generation set points to EV CSO, Household Coordinator, Storage, DER, Substations;
8. The DMS updates periodically Ancillary Services with supporting information (e.g. extreme increases or decreases in electricity frequency);
9. Ancillary services forward the info to the Transmission System Operator (TSO);
10. The TSO provides info to the Market for setting the energy price;
11. The price of the energy can be requested by aggregators and other energy dealers (e.g. E-mobility);
12. The Market provides energy price to the Aggregator, that forwards the price to the energy dealers and to controllers (i.e. CSO, Coordinator, EMG),
13. Information about the demand and the generation Flexibilities are provided to the aggregator by Storages, DERs, CSO and EMG.

Viewpoint Examination. Consider now the case where the owner of the HH has the capability of storing energy from DER and also owns an EV. As described in Sect. 2.1,

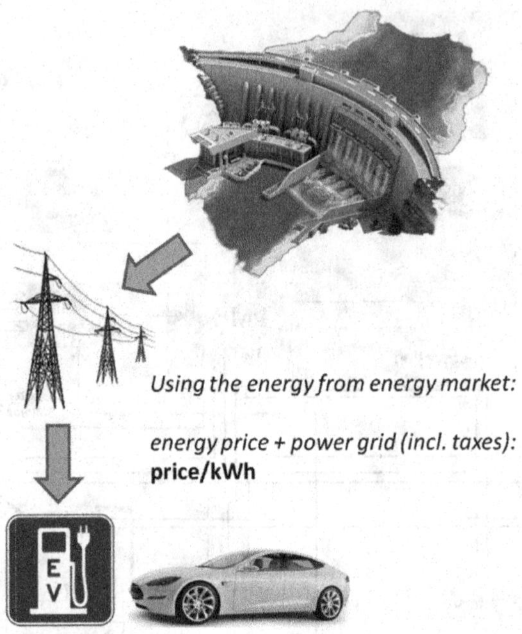

Fig. 23. The energy used to recharge an EV is bought from the market

their EV can be charged at a CSO by means of a CP, the energy used in this process is generally bought by the CSO from the energy market and sold at a specific price per each kWh (plus power grid fees including taxes) (Fig. 23).

The CSO may also buy energy directly from the HHs for recharging EVs, like, e.g., when the HH energy price is cheaper than the one provided by the market or for compensating effects of some energy-related service disruptions (e.g., disconnected energy generators). In this case the total energy price would be the energy cost from the HH plus the power grid costs with taxes (Fig. 24).

From the above scenario, the below beneficial emergence scenarios were identified:

Consider now the scenario in which the owner of the HH wants to recharge its EV to a CSO. It is evident that the energy needed to recharge the EV can be bought and provided, by the CSO, form either the energy market or HHs, including the one owned by the driver of the EV.

The HH owner can check the energy price, relative to the energy market, of the CSO and decide whether to recharge its EV with such energy or to use the energy stored at home using DER, i.e. the HH owner can compare the price of the energy stored at home with the one sold by the market and decide accordingly (Fig. 26).

Suppose that the energy on the market is more expensive than the one stored at the HH. The HH owner will then decide to use its own stored energy to recharge its EV, enabling the energy transmission from the house to the CSO. Using the energy stored at home will cost the owner to pay only the provision of power grid including taxes (the price may be related to the total power grid usage time or to total transmitted kWh) (see Figs. 25 and 27).

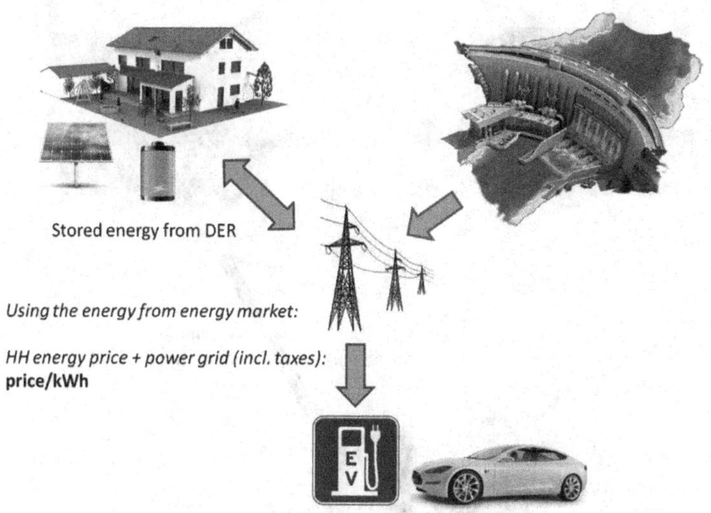

Fig. 24. The CSO buys energy from the HH.

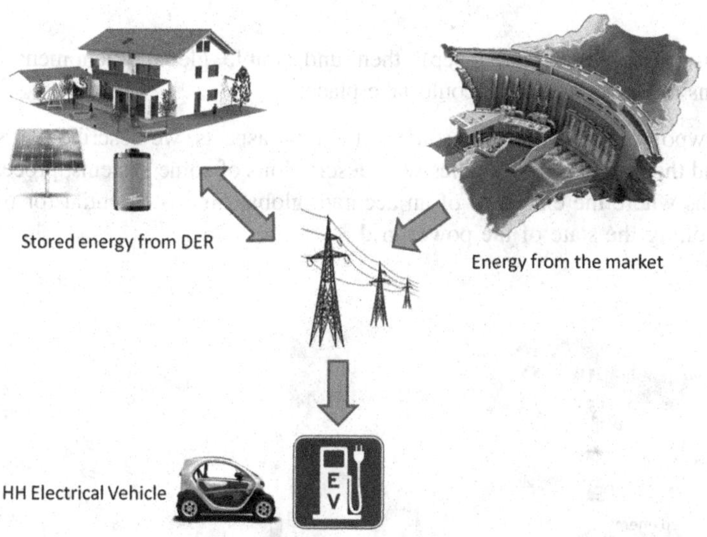

Fig. 25. The CSO can buy energy from either the market or HHs

These new emergent behaviors clearly arise with the interconnection and the available communications between smart constituent system of a SoS, which the AMADEOS framework is able to capture and describe. This way the possible large variations in the network load could be attenuated by a suitable adaptation of the prices. However, if this adaptation process is not correctly implemented (for instance, if the

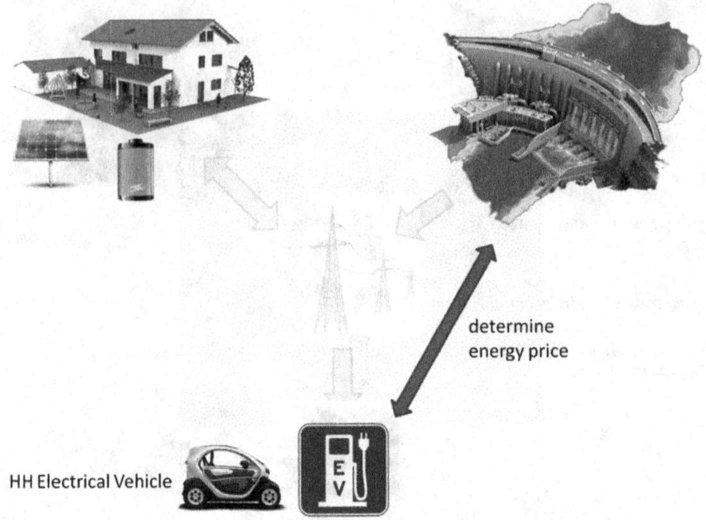

Fig. 26. The HH owner can check and compare the price of the energy stored at home with the one of the market

adaptation is too quick and steep), then undesirable global phenomena, such as oscillations of prices and load, could take place.

Time Viewpoint. For identifying some of the time aspects, we describe two scenarios that extend the previous case studies with descriptions of some systems, processes, and interactions where the existence of an accurate global time is essential for measuring and controlling the state of the power grid.

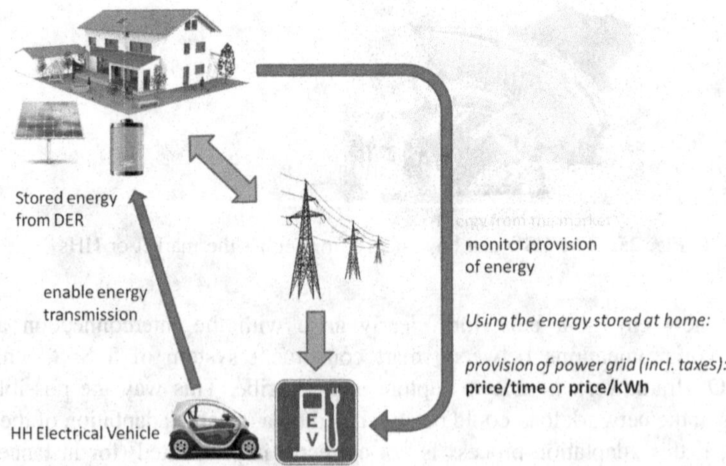

Fig. 27. The HH owner decide to recharge its EV with the energy stored at home

Scenario 1: Maintaining the Power Phasor Parameters. The power phasor is defined by the amplitude, frequency, and relative phase of the electric current or voltage in a power line. As the load in the power network varies, the braking momentum of the power generators increases, so the generators' rotors slow down. All variations in network load lead thus to variations in power frequency. Traditionally, these variations were compensated locally, by applying the load measurement signals to an Automated Generator Control (AGC), which basically is an electro-mechanical governor that compensates for the changes in braking momentum. The AGC signals are collected from different points in the network, and forwarded to large production facilities [5]. In order for the AGC to work efficiently, the following requirements must be satisfied:

- Timely measurements of network loads.
- Timely adjustment of voltage controllers.

The adjustment signals for AGC are generated once every 2 to 4 s, so the transmission time requirements are not very strict [5].

Another phenomenon that alters the power phasor is the reactive power produced by the reactance (i.e. the reactive part of impedance) of a load. A capacitive load will make the voltage lag behind the current, whereas an inductive load will make the current lag behind the voltage. Although the reactive power does not dissipate energy, it affects the efficiency of power transmission to the consumers, so this too must be controlled, such that the reactive angle φ_X is kept as close as possible to zero.

To compensate for reactive power introduced in the network primarily by motors and transformers, capacitive loads (capacitor banks) can be switched on or off by control units. Unlike the AGC, which acts centrally at the production facility, the reactive power control can be distributed across the grid [6]. In the AMADEOS scenarios this can be achieved by the LMO.

In a purely reactive approach the individual measurements by SMs are collected by LMOs (via the EMGs or CSOs), which try to compensate for load variations using the locally available power reserves and by re-scheduling, when possible, of some loads. When these local measures are not sufficient, appropriate AGC signals are transmitted to the power generation facilities. Of course, particular implementations could include additional aggregation layers, but those do not change significantly the problem.

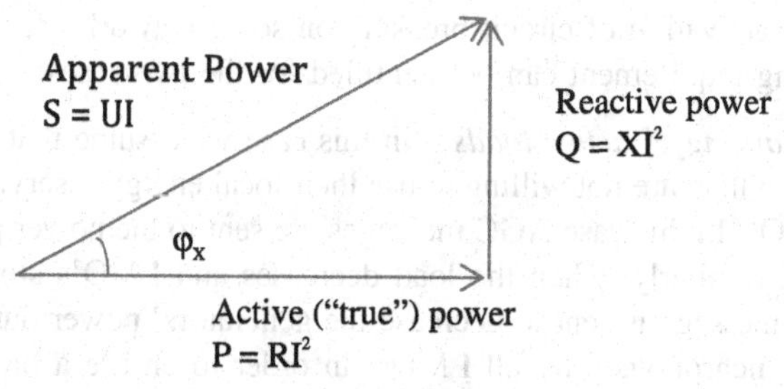

Fig. 28. Power triangle

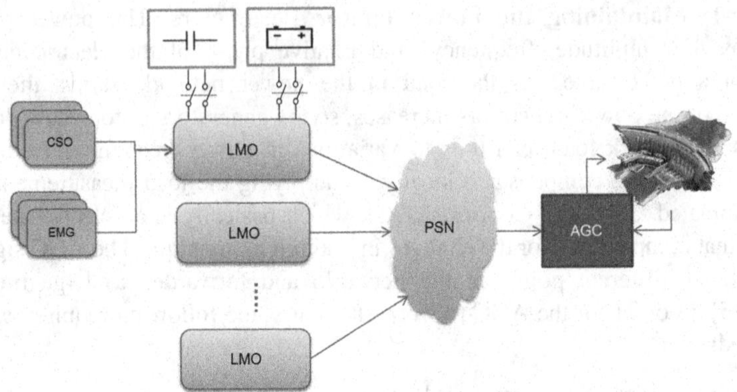

Fig. 29. Load measurement and control architecture

Concerning the reactive power in the grid, the LMOs try to locally compensate. In case when the local capacitor banks do not suffice, a signal must be broadcasted (Reactive Load Control - RLC) to request additional capacitive loads be switched on by other LMOs.

Overall, the load measurement and control architecture includes a number of geographically distributed LMOs and one local AGC for each of the power plants (see Fig. 28).

The effectiveness of this reactive approach is limited, due to the delays in the transmission of AGC signals and the time required for adjustments. A more advanced approach (see Fig. 29) tries to predict the variation of the load based on SCADA measurements from various places in the network, and a forecasting of load variations [7]. The prediction function is based on a sampling of the real load over some past time interval.

Although the AGC and reactive load balancing are still slow (in the order of seconds), a good estimation of the current state of the system, as well as the load prediction, require better time accuracy than in the previous case, and global time awareness. Indeed, if two consecutive measurements from a given location in the power grid arrive in reverse order, then the detected variation trend is reversed. If this reversal occurs randomly over a large number of measurements, then the whole predictive load balancing has, on average, no effect. In a worst-case scenario the order of a sequence of measurements can be altered such that the variations are amplified, leading potentially to the activation of circuit breakers on some network segments.

Specific timing requirement can be identified for the two cases:

A. *Predictive balancing* of *active loads* – in this case we assume that LMOs compete with each other, so they are not willing to use their local energy reserves to balance the load at other LMOs. In this case AGC messages are sent to the power plant to increase its power output. Similarly, when the load decreases and LMO's storage capacity is reached, another message is sent to decrease the generators' power output. Obviously, these are sent asynchronously by all LMOs. In order to ensure a proper ordering of these events at the local AGC Message Queue, all AGC messages must be

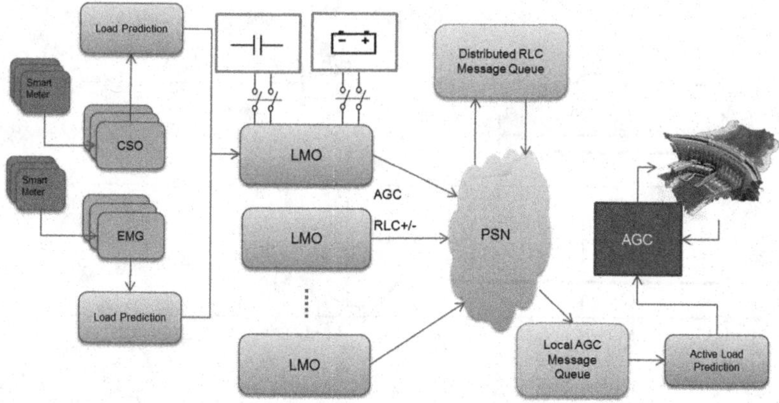

Fig. 30. Load measurement and predictive control architecture

time-stamped. The time difference between the local clocks that provide the time stamps must be smaller than the sampling interval[2] of the Active Load Prediction (ALP) function.

B. Predictive balancing of *reactive loads* – in this case we assume that LMOs help each other since the cost of switching capacitor banks is low. A simple protocol could be defined as follows:

- When a LMO needs additional capacitive reactance it broadcasts a RLC request message (RLC+), and when this is no longer needed it broadcasts a RLC−.
- All RLC messages from different LMOs get stored in the Distributed RLC Queue.
- All LMOs poll the queue and:
  - If the first message is RLC+ and the polling LMO has the disposable capacitance requested by that message, it removes this message in the queue and then switches on an appropriate capacitance.
  - If the first message is RLC− and the polling LMO has more allocated capacitance than that requested by the message, it removes this message and switches off the appropriate capacitance.

Of course, more elaborated and efficient protocols can be implemented for processing more messages at once, but for the purposes of this study this is sufficient.

Since each LMO has a local load predictor, the polling of the queue needs to be done at the same rate as the same sampling rate of the load prediction function. For a correct functioning of these predictors, the messages must be correctly ordered based on time stamps, as in the case of AGC. Additionally, the polling cannot be done completely asynchronously since this could lead to race conditions due to the distributed queue.

---

[2] Note that when the guaranteed transmission time for the AGC messages is smaller than the ALP sampling interval, then no time stamping (and thus no global time) is needed.

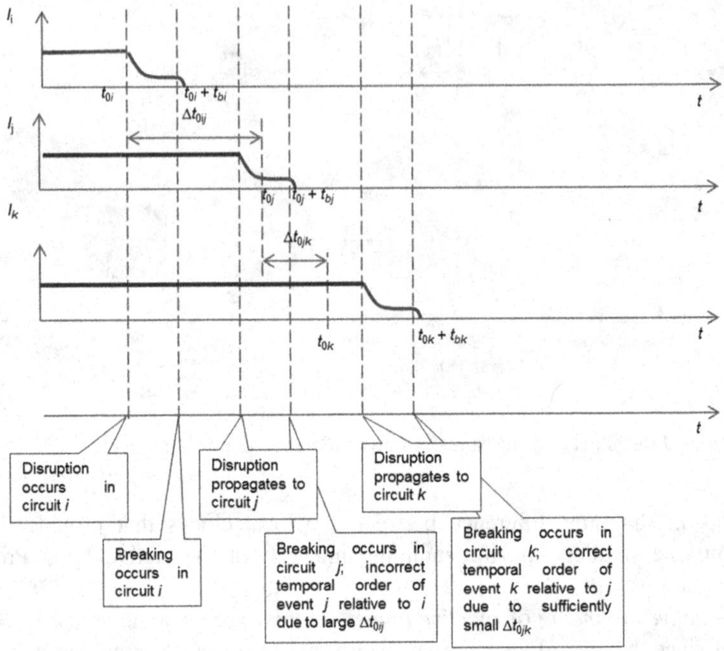

Fig. 31. Cascaded propagation of disruption and time stamping of breaking events; in nominal terms, $t_{0i} = t_{0j} = t_{0k}$

The time requirements resulting from these two cases are:

1. The maximum synchronisation error should be smaller than the sampling interval of the load prediction function; otherwise, two messages arriving in reversed order could indicate a wrong trend (cases A and B)
2. The polling of the queue should be done at specific time slots allocated to each LMO (time-division multiplex); the maximum synchronisation error (i.e. difference between local clocks) should be smaller than the time slot minus the time required for processing the queue (case B)

Scenario 2: Forensic Analysis of Disruption Events. When a disruption occurs in a network, it may lead to cascading effects (Fig. 31). In case of complex networks of producers and consumers it is not always easy to identify the original event that initiated the cascaded failures. However, this information is crucial when network design needs to be improved to increase the resilience of the network, or when the network operators are liable for the quality of the services they provide.

To cope with this requirement each circuit breaking event could be logged by the breakers' control computer. This way a chain of events can be reconstructed and the root cause identified. However, given the high speed in which these events occur[3], the temporal order of the logged events may be distorted by the differences in the local

[3] A typical reaction time for modern equipment is 1/8th of a 20 ms cycle [8].

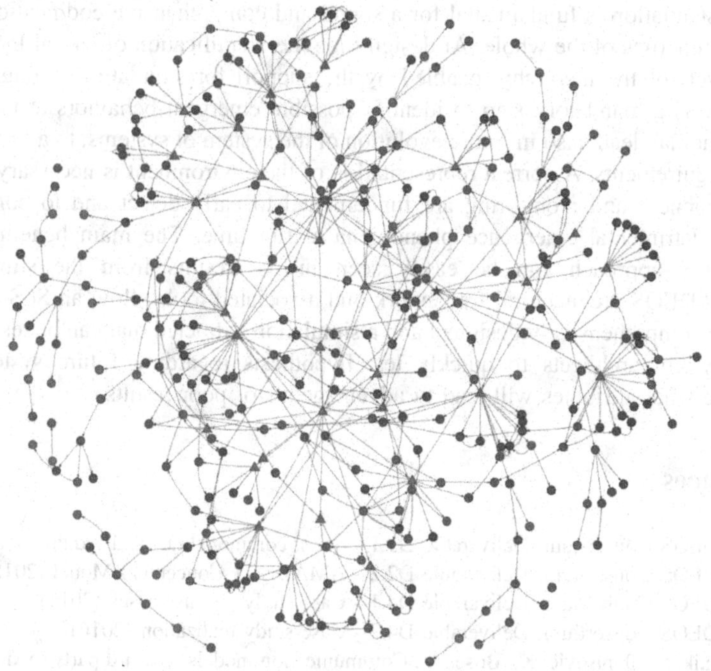

Fig. 32. The full SmartGrid model generated using the model query: "return true;" (i.e.: select all blocks). Here, the triangles represent systems, star represents the SoS and other blocks are represented by circles. The color of the blocks is the same as the one in the Blockly model.

clocks used for time stamping. For this reason, it is necessary that the synchronization error between local clocks is kept below the breaking time of the fastest breaker plus the propagation time of the disruption. This is illustrated in Fig. 30, where a time value t_0 corresponds to three different moments in three circuits whose local clocks are not synchronized. If the delay between two local clocks is too large, as it is the case for circuits i and j, then the events cannot be ordered correctly.

Finally, for both scenarios the maximum transmission times for all the messages related to events occurring in the smart grid must be guaranteed. If this is not the case, then most of the functions described here cannot be implemented.

3 Conclusion

This chapter has shown how a realistic case study can be modeled in the Blockly tool and directly simulated. A high-level view of the entire model can be seen using the model query tool of supporting facility to select all blocks. Below is the graph of the model consisting of elements and relationship between each blocks. The diagram shows the complexity of the full model (Fig. 32).

Modeling complex and pervasive infrastructures as the one used as case study clearly highlights how the support of a precise conceptual model and of specific tools for its instantiation is fundamental for a sound and comprehensive codification of the various properties of the whole. At design time the identification of causal loop in the lower levels of the hierarchy, enabled by the support for simulation through model execution, is a mandatory step to identify possible emergent behaviors at the higher levels, that may lead, also in future evolution of the system of systems, to a violation of system requirements. A correct representation of the environment is necessary. Global time Awareness and monitoring are fundamental to early detect and to contain the effect of detrimental emergence phenomena at run time. The main benefits of the AMADEOS approach can be easily seen in the results from the simulations: The AMADEOS architectural framework and associated tools allow an SoS architecture to be comprehensively designed and a simulation extracted that can be tested. This allows system architects to quickly test hypothesis regarding future systems and determine what attributes will lead to advantageous or poor results.

References

1. AMADEOS Consortium, Deliverable D1.1 - SoSs, commonalities and requirements (2014)
2. AMADEOS Consortium, Deliverable D2.2 - AMADEOS Conceptual Model (2015)
3. AMADEOS Consortium, Deliverable D4.1 - Case study and use cases (2015)
4. AMADEOS Consortium, Deliverable D4.2 - Case study realization (2016)
5. Bhowmik, S., Tomsovic, K., Bose, A.: Communication models for third party load frequency control. IEEE Trans. Power Syst. **19**(1), 543–548 (2004)
6. Saypaserth, P., Premrudeepreechacharn, S.: Allocating reactive power using artificial neural network. In: 2011 Asia-Pacific Power and Energy Engineering Conference (APPEEC), Wuhan, pp. 1–4 (2011). doi:10.1109/APPEEC.2011.5749101
7. Tomsovic, K., Bakken, D.E., Venkatasubramanian, V., Bose, A.: Designing the next generation of real-time control, communication, and computations for large power systems. Proc. IEEE **93**(5), 965–979 (2005)
8. ABB Miniature Circuit Breakers – Application Guide. https://library.e.abb.com/public/ef5f3a6e43c99c288525758500534fe6/1SXU400142M0201.pdf

Glossary

Absolute Timestamp
An absolute timestamp of an event is the timestamp of this event that is generated by the reference clock

Acceptance Test
A test that determines if a state in the problem space is a member of the solution set

Access Control
Access control is concerned with providing control over security critical actions that take place in a system. Providing control over actions consists of explicitly determining either the actions that are permitted by the system, or explicitly determining the actions that are not permitted by the system

Access Control Model
An access control model captures the set of allowed actions as a policy within a system

Accuracy
The *accuracy of a clock* denotes the maximum offset of a given clock from the external time reference during the IoD, measured by the reference clock

Acknowledged SoS
Independent ownership of the CSs, but cooperative agreements among the owners to an aligned purpose

Action
The execution of a program by a computer or a protocol by a communication system

Action Sequence
A sequence of actions, where the end-signal of a preceding action acts as the start signal of a following action

Activity Interval
The interval between the start signal and the end signal of an action or a sequence of related actions

Actuator
An actuator is an interface device that accepts data and control information from an interface component and realizes the intended physical effect at its placement in the physical environment

Agility (of a system)
Quality metric that represents the ability of a system to efficiently implement evolutionary changes

Architectural Style
The set of explicit or implicit rules and conventions that determine the structure and representation of the internals of a system, its data and protocols

Arrival Instant
The instant when the first bit of a message arrives at the receiver

Artifact
An entity that has been intentionally produced by a human for a certain purpose

Atomic Action
An atomic action is an action that has the all-or-nothing property. It either completes and delivers the intended

result or does not have any effect on its environment

Attribute
A characteristic quality of an entity

Authentication
The process of verifying the identity or other attributes claimed by or assumed of a subject, or to verify the source and integrity of data

Authorization
Authorization is the mechanism of applying access rights to a subject. Authorizing a subject is typically processed by granting access rights to them within the access control policy

Authority
The relationship in which one party has the right to demand changes in the behavior or configuration of another party, which is obliged to conform to these demands

Autonomous System
A system that can provide its services without guidance by another system

Availability
Readiness for service

Behavior
The timed sequence of the effects of input and output actions that can be observed at an interface of a system

Bottom Up Design
A hierarchical design methodology where the design starts at the bottom of the holarchy or formal hierarchy

Business Value
Overarching concept to denote the performance, impact, usefulness, etc. of the functioning of the SoS

Capability
Ability to perform a service or function

Cascade Effect
A cascade effect exists, if in a system with a multitude of parts at the micro level a state change of a part at the micro-level causes successive state changes of many other parts at the micro level such that the cumulative effect of the totality of these state changes results in a novel phenomenon

Causal Loop
A causal loop exists, if the emergent property at the macro-level causes a change of the state of the parts at the micro-level

Causal Model
Abstract model describing the causal dependencies between relevant variables in a given domain

Causal Order
A causal order among a set of events is an order that reflects the cause-effect relationships among the events

Channel
A logical or physical link that transports information among systems at their connected interfaces

Channel Model
A model that describes effects of the channel on the transferred information

Checked Message
A message is checked at the source (or, in short, checked) if it passes the output assertion

Ciphertext
Data in its encrypted form

Clock
A (digital) clock is an autonomous system that consists of an oscillator and a register. Whenever the oscillator completes a period, an event is generated that increments the register

Clock Ensemble
A collection of clocks, not necessary in the same physical location, operated together in a coordinated way either for mutual control of their individual properties or to maximize the performance (time accuracy and frequency stability) and availability of a time-scale derived from the ensemble)

Collaborative SoS
Voluntary interactions of independent CSs to achieve a goal that is beneficial to the individual CS

Communication Action
An action that is characterized by the execution of a communication protocol by a communication system

Communication Protocol
The set of rules that govern a communication action

Compatibility (full, Itom)
The Itom that is sent by the sender is received by the receiver without modification

Component
A subsystem of a system, the internal structure of which is of no interest

Computational Action
An action that is characterized by the execution of a program by a machine

Concept
A category that is augmented by a set of beliefs about its relations to other categories, i.e., existing knowledge, is called a concept

Concise State
The state of a system is considered concise if the size of the declared ground state is at most in the same order of magnitude as the size of the system's largest input message

Confidentiality
The absence of unauthorized disclosure of information

Configuration Interface (C-Interface)
An interface of a CS that is used for the integration of the CS into an SoS and the reconfiguration of the CS's RUIs while integrated in a SoS

Connected Interfacc
An interface that is connected to at least one other interface by a channel

Connection System/Gateway Component/Wrapper
A new system with at least two interfaces that is introduced between interfaces of the connected component systems in order to resolve property mismatches among these systems (which will typically be legacy systems)

Consistency
The property of a set of entities that see the same data at the same time

Constituent System (CS)
An autonomous subsystem of an SoS, consisting of computer systems and possibly of controlled objects and/or human role players that interact to provide a given service

Constraint
A restriction in the problem space

Construct
A non-physical entity, a product of the human mind

Consume/Produce (CP) Paradigm
At the sender, the communication system consumes the message from a sender queue and at the receiver the communication system adds the received message to a receiver queue

Context
The set of cultural circumstances, conventions or facts, and the time that surround and have a possible influence on a particular thing, construct, event, situation, system, etc. in the UoD

Context Compatibility
the same data (bit pattern) is explained in the same way at the sender and at the receiver

Context Incompatibility
the same data (bit pattern) is explained differently at the sender and at the receiver

Contract
Agreement between two or more parties, where one is the customer and the others are service providers. This can be a legally binding formal or an informal "contract". It can be expressed in terms of objectives

Control Flow
The flow of control signals when executing a protocol

Coordinated Clock
A clock synchronized within stated limits to a reference clock that is spatially separated

Correct Message
A message is correct if it is both timely and value correct

Critical Service
A critical service is the service of a system that requires a specific criticality level

Criticality
Criticality is a designation of the required criticality level for a system component

Criticality Level
The criticality level is the level of assurance against failure

Cryptography
The art and science of keeping data secure

Cyber-Physical System (CPS)
A system consisting of a computer system (the cyber system), a controlled object (a physical system) and possibly of interacting humans

Cyber Space
Cyber space is an abstraction of the Universe of Discourse (UoD) that consists only of information processing systems and cyber channels to realize message-based interactions

Cycle
A temporal sequence of significant events that, upon completion, arrives at a final state that is related to the initial state, from which the temporal sequence of significant events can be started again

Data
A data item is an artefact, a pattern, created for a specified purpose

Data Flow
The flow of the payload data of a message from a sender to the receivers

Datagram
A best effort message transport service for the transmission of sporadic messages from a sender to one or many receivers

Declared Ground State
A declared data structure that contains the relevant ground state of a given application at the ground state instant

Decryption
The process of turning ciphertext back into plaintext

Dependability
The ability to deliver service that can justifiably be trusted

Design
The process of defining an architecture, components, modules and interfaces of a system to satisfy specified requirement

Design for Evolution
Exploration of forward compatible system architectures, i.e. designing applications that can evolve with an ever-changing environment. Principles of evolvability include modularity, updateability and extensibility. Design for evolution aims to achieve robust and/or flexible architectures

Design for Evolution in the context of SoSs
Design for evolution means that we understand the user environment and design a large SoS in such a way that expected changes can be accommodated without any global impact on the architecture. 'Expected' refers to the

fact that changes will happen, it does not mean that these changes themselves are foreseeable

Design for Testability
The architectural and design decisions in order to enable to easily and effectively test our system

Design Inspection
Examination of the design and determination of its conformity with specific requirements

Design Walkthrough
Quality practice where the design is validated through peer review

Designer
An entity that specifies the structural and behavioral properties of a design object

Deterministic Behavior
A system behaves deterministically if, given an initial state at a defined instant and a set of future timed inputs, the future states, the values and instants of all future outputs are entailed

Diagnosis Interface (D-Interface)
An interface that exposes the internals of a Constituent System (CS) for the purpose of diagnosis

Directed SoS
An SoS with a central managed purpose and central ownership of all CSs

Downward Causation
The phenomenon that some novel macro-level properties have causal powers to control the micro-level properties from which they emerge

Drift
The drift of a physical clock is a quality measure describing the frequency ratio between the physical clock and the reference clock

Duration
The length of an interval

Dynamicity of a system
The capability of a system to react promptly to changes in the environment

Emergence
A phenomenon of a whole at the macro-level is emergent if and only if it is of a new kind with respect to the non-relational phenomena of any of its proper parts at the micro level

Encryption
The process of disguising data in such a way as to hide the information it contains

End Signal
An event that is produced by the termination of an action

Entity
Something that exists as a distinct and self-contained unit

Entourage of a CPS
The entourage is composed of those entities of a CPS (e.g., the role playing human, controlled object) that are external to the cyber system of the CPS but are considered an integral part of the CPS

Environment of a System
The entities and their actions in the UoD that are not part of a system but have the capability to interact with the system

Environmental Dynamics
Autonomous environmental processes that cause a change of state variables in the physical environment

Environmental Model
A model that describes the behavior of the environment that is relevant for the interfacing entities at a suitable level of abstraction

Epoch
An instant on the timeline chosen as the origin for time-measurement

Error
Part of the system state that deviated from the intended system state and could lead to system failure

Error Containment
Error Containment prevents propagation of errors by employing error detection and a mitigation strategy

Error Containment Region (ECR)
A set of at least two Fault Containment Regions (FCRs) that perform error containment

Established Rule
An observed consequence that often follows if a set of antecedent conditions applies

Event
A happening at an instant

Event Variable
A variable that holds information about some change of state at an instant

Event-triggered (ET) Action
An action where the start signal is derived from an event other than the progression of time

Evolution
Process of gradual and progressive change or development, resulting from changes in its environment (primary) or in itself (secondary)

Evolutionary Performance
A quality metric that quantifies the business value and the agility of a system

Evolutionary Step
An evolutionary change of limited scope

Evolvable architecture
An architecture that is adaptable and then is able to incorporate known and unknown changes in the environment or in itself

Execution Time
The duration it takes to execute a specific action on a given computer

Explained Emergence
An emergent phenomenon that is observed at a macro level is explained emergent if a trans-ordinal law that explains the occurrence of the emergent phenomenon at the macro level out of the properties and interactions of the parts at the adjacent micro level is known (or has been formulated post facto)

Explanation
The explanation of the data establishes the links between data and already existing concepts in the mind of a human receiver or the rules for handling the data by a machine

Explicit Flow Control
After having sent a message, the sender receives a control message from the receiver informing the sender that the receiver has processed the sent message

External Clock Synchronization
The synchronization of a clock with an external time base such as GPS

External Interface
A Constituent System (CS) is embedded in the physical environment by its external interfaces

Failure
The actual system behavior deviation from the intended system behavior

Failure Modes
The forms that the deviations from the system service may assume; failure modes are ranked according to failure severities (e.g. minor vs. catastrophic failures)

Fault
The adjudged or hypothesized cause of an error; a fault is active when it causes an error, otherwise it is dormant

Fault Containment Region (FCR)
A Fault Containment Region (FCR) is a collection of components that

operates correctly regardless of any arbitrary fault outside the region

Fault Forecasting
The means to estimate the present number, the future incidence, and the likely consequences of faults

Fault Prevention
The means to prevent the occurrence or introduction of faults

Fault Removal
The means to reduce the number and severity of faults

Fault Tolerance
The means to avoid service failures in the presence of faults

Flexible Architecture
Architecture that can be easily adapted to a variety of future possible developments

Flow Control
The control of the flow of messages from the sender to the receiver such that the sender does not outpace the receiver

Formal Problem
A problem in a well-defined problem space

Frequency Drift
A systematic undesired change in frequency of an oscillator over time

Frequency Offset
The frequency difference between a frequency value and the reference frequency value

Function
A function is a mapping of input data to output data

Gateway
A transformation system in cyberspace

General Law
An inevitable consequence that follows if a set of antecedent conditions applies

Global Evolution
Global evolution affects the SoS service and thus how CSs interact. Consequently, global evolution is realized by changes to the Relied Upon Interface (RUI) specifications

Governance
Theoretical concept referring to the actions and processes by which stable practices and organizations arise and persist. These actions and processes may operate in formal and informal organizations of any size; and they may function for any purpose

Global Positioning System Disciplined Oscillator (GPSDO)
The GPSDO synchronizes its time signals with the information received from a GPS receiver

Granularity/Granule of a Clock
The duration between two successive ticks of a clock is called the granularity of the clock or a granule of time

Ground State
At a given level of abstraction, the ground state of a cyclic system is a state at an instant when the size of the instantaneous state space is at a minimum relative to the sizes of the instantaneous state spaces at all other instants of the cycle

Ground State Instant
The instant of the ground state in a cyclic system

Hierarchical Design
A design methodology where the envisioned system is intended to form a holarchy or formal hierarchy

Holarchy
A structure where holons at one level interact horizontally to form a novel holon at the next higher level

Holdover
The duration during which the local clock can maintain the required precision of the time without any input from the GPS

Holon
A two-faced entity in a non-formal hierarchy that acts externally at the macro-level as a whole while it is established internally by the interactions of its parts at the micro-level

Homogenous System
A system where all sub-systems adhere to the same architectural style

Human-Machine Interface (HMI) Component
A component of the CS that realizes the human-machine interface of a CS

Idempotent Action
An action is idempotent if the effect of executing it more than once has the same effect as of executing it only once

Implicit Flow Control
The sender and receiver agree a priori on a maximum send rate. The sender commits to never send messages faster than the agreed send rate and the receiver commits to accept all messages that the sender has sent

Incentive
Some motivation (e.g., reward, punishment) that induces action

Information
A proposition about the state of or an action in the world

Initial State
(i) an existing deficient state of affairs that needs a solution or (ii) a recognized opportunity that should be exploited or (iii) a formal statement of a question (academic story problem)

Input Action
An action that reads or consumes input data at an interface

Input Data
Data that is used as an input to a system

Insidious Message
A message is insidious if it is permitted but incorrect

Instant
A cut of the timeline

Instantaneous State Space
The state space of a system is formed by the totality of all possible values of the state variables at a given instant

Integrity
The absence of improper system state alterations

Interaction
An interaction is an exchange of information items at connected interfaces

Interface
A point of interaction of a system with another system or with the system environment

Interface Physical Specification (P-Spec)
Part of the CP-Spec that concerns the specification of exchanges with the physical environmental model

Interface Cyber-Physical Specification (CP-Spec)
Part of the interface specification that concerns interface properties at the cyber-physical interface layer

Interface Itom Specification (I-Spec)
Part of the interface specification that concerns interface properties at the informational interface layer

Interface Layer
An abstraction level under which interface properties can be discussed

Interface Message Specification (M-Spec)
Part of the CP-Spec that concerns the specification of messages exchanged with the cyber space environmental model

Interface Model
The interface model contains the explanation of the data sent or received over this interface and thus establishes the Itoms

Interface Properties
The valued attributes associated with an interface

Interface Service Specification (S-Spec)
Part of the interface specification that concerns interface properties at the service interface layer

Interface Specification
The interface specification defines at all appropriate interface layers the interface properties, i.e., what type of, how, and for what purpose information is exchanged at that interface

Internal Clock Synchronization
The process of mutual synchronization of an ensemble of *clocks* in order to establish a *global time* with a bounded *precision*

Internal Interface
An interface among two or more subsystems of a Constituent System (CS)

Interval
A section of the timeline between two instants

Interval of Discourse (IoD)
The Interval of Discourse specifies the time interval that is of interest when dealing with the selected view of the world

Intra-ordinal Law
A new law that deals with the emerging phenomena at the macro level

Irrevocable Action
An action that cannot be undone

Itom
An Itom (Information Atom) is a tuple consisting of data and the associated explanation of the data

Jitter
The short-term phase variations of the significant instants of a timing signal from their ideal position on the time-line (where long-term implies here that these variation of frequency are greater than or equal to 10 Hz) (see also wander)

Jitter of a Message
The duration between the minimal transport duration and the maximum transport duration

Key
A numerical value used to control cryptographic operations, such as decryption and encryption

Legacy System
An existing operational system within an organization that provides an indispensable service to the organization

Local Evolution
Local evolution only affects the internals of a Constituent System (CS) which still provides its service according to the same and unmodified Relied Upon Interface (RUI) specification

Local I/O Interface (L-Interface)
An interface that allows a Constituent System (CS) to interact with its surrounding physical reality that is not accessible over any other external interface

Maintainability
The ability to undergo modifications and repairs

Managed Evolution
Evolution that is guided and supported to achieve a certain goal

Managed SoS Evolution
Process of modifying the SoS to keep it relevant in face of an ever-changing environment

Meet-in-the-Middle Design
A hierarchical design methodology where the top down design and the bottom up design are intermingled

Message
A data structure that is formed for the purpose of the timely exchange of information among computer systems

Message Variable
A tuple consisting of a syntactic unit of a message and a name, where the name points to the explanation of the syntactic unit

Message-based Interface Port
The message-based interface contains ports (i.e., channel endpoints) where message payloads can be placed for sending, or received message payloads can be read from

Meta Data
Data that describes the meaning of object data

Metric
Indicator used to quantitatively describe an attribute of the system, like throughput for performance or availability for dependability

Major Evolutionary Step
An evolutionary step that affects the Relied Upon Interface (RUI) Itom specification and might need to be considered in the management of SoS dynamicity and SoS emergence

Minor Evolutionary Step
An evolutionary step that does not affect the Relied Upon Interface (RUI) Itom Specification (I-Spec) and consequently has no effects on SoS dynamicity or SoS emergence

Modularity
Engineering technique that builds larger systems by integrating modules

Module
A set of standardized parts or independent units that can be used to construct a more complex structure

Monolithic System
A system is called monolithic if distinguishable services are not clearly separated in the implementation but are interwoven

Multi-Criteria Decision Analysis (MCDA)
MCDA is a sub-discipline of operations research that explicitly considers multiple criteria in decision-making, allowing the evaluation of one or more decision alternatives in light of the multiple criteria

Multi-criticality System
A multi-criticality system has at least two components that have a different criticality

Nominal Frequency
The desired frequency of an oscillator

Non-Sparse Events
Events that occur in the passive interval of the sparse time

Now
The instant that separates the past from the future

Object
Passive system-related devices, files, records, tables, processes, programs, or domain containing or receiving information. Access to an object implies access to the information it contains

Object Data
Data that is the object of description by meta data

Objective
Values for the quality metrics to be attained

Observation of an Entity
An atomic structure consisting of the name of the entity, the name and the value of the attribute (i.e., the property), and the timestamp denoting the instant of observation

Offset of events
The offset of two events denotes the duration between two events and the position of the second event with respect to the first event on the timeline

Open System
A system that is interacting with its environment during the given IoD

Output Action
An action that writes or produces output data at an interface

Output Data
Data that is produced by a system

PAR-Message
A PAR-Message (Positive Acknowledgment or Retransmission) is an error controlled transport service for the transmission of sporadic messages from a sender to a single receiver

Payload of a Message
The bit pattern carried in the data field of the message

Period
A cycle marked by a constant duration between the related states at the start and the end of the cycle

Periodic System
A system where the temporal behavior is structured into a sequence of periods

Permission
Attributes that specify the access that subjects have to objects in the system

Permitted Message
A message is permitted with respect to a receiver if it passes the input assertion of that receiver. The input assertion should verify, at least, that the message is valid

Phase
A measure that increases linearly in each period from 0 degrees at the start until 360 degrees at the end of the period

Phase alignment
The alignment of the phases between two periodic systems exhibiting the same period, such that a constant offset between the phases of the two systems is maintained

Plaintext
Unencrypted data

Precision
The precision of an ensemble of synchronized clocks denotes the maximum offset of respective ticks of the global time of any two clocks of the ensemble over the IoD. The precision is expressed in the number of ticks of the reference clock

Primary Clock
A clock whose rate corresponds to the adopted definition of the second. The primary clock achieves its specified accuracy independently of calibration

Prime Mover
A human that interacts with the system according to his/her own goal

Private Key
In an asymmetric cryptography scheme, the private or secret key of a key pair which must be kept confidential and is used to decrypt messages encrypted with the public key

Problem
A perceived need to transform an initial state to a goal state

Property
A valued attribute

Property Mismatch
A disagreement among connected interfaces in one or more of their interface properties

Public Key
A cryptographic key that may be widely published and is used to enable the operation of an asymmetric cryptography scheme. This key is mathematically linked with a corresponding private key

Public Key Cryptography
Cryptography that uses a public-private key pair for encryption and decryption

Quality
The standard of something as measured against other things; the degree of excellence of something

Quality of Service
The ability of a system to meet certain requirements for different aspects of the system like performance, dependability, evolvability, security or cost; possibly expressed in terms of levels and quantitatively evaluated through metrics

Raw Data
The bit pattern that is produced by a sensor system

Read/Write (RW) Paradigm
At the sender the communication system reads the contents of the message from a message variable and at the receiver the communication system writes the arriving message into a message variable, overwriting the old content of the message variable

Real-Time (RT) Transaction
A transaction that must complete before a specified deadline

Real-Time System (RTS)
A computer system for which the correct results must be produced within time constraints

Reasonableness Condition
The reasonableness condition of clock synchronization states that the granularity of the global time must be larger than the precision of the ensemble of clocks

Receive Instant
The instant when the last bit of a message arrives at the receiver

Reconfigurability
The capability of a system to adapt its internal structure in order to mitigate internal failures or to improve the service quality

Reducible System
A system where the sum of the parts makes the whole

Reference Clock
A hypothetical clock of a granularity smaller than any duration of interest and whose state is in agreement with TAI

Reference Monitor
A reference monitor represents the mechanism that implements the access control model. A reference monitor is defined as: An access control concept that refers to an abstract machine that mediates all accesses to objects by subjects

Refined Data
Data that has been created by a purposeful process from the raw data to simplify the explanation of the data in a given context

Reliability
Continuity of service

Relied upon Interface (RUI)
An interface of a CS where the services of the CS are offered to other CSs

Relied upon Message Interface (RUMI)
A message interface where the services of a CS are offered to the other CSs of an SoS

Relied upon Physical Interface (RUPI)
A physical interface where things or energy are exchanged among the CSs of an SoS

Relied upon Service (RUS)
(Part of) a Constituent System (CS) service that is offered at the Relied Upon Interface (RUI) of a service providing CS under a Service Level Agreement (SLA)

Requirement
A statement that identifies a necessary attribute, capability, characteristic, or quality of a system

Reservation
A commitment by a service provider that a resource that has been allocated to a service requester at the reservation allocation instant will remain allocated until the reservation end instant

Reservation Allocation Instant
The instant when a resource

reservation is allocated to a service requestor by a service provider

Reservation End Instant
The instant until a reservation is allocated to a service provider

Reservation Request Instant
The instant when a resource is requested by a service requestor

Resultant Phenomenon
A phenomenon at the macro-level is resultant if it can be reduced to a sum of phenomena at the micro-level

Risk
A measure of the extent to which an organization is threatened by a potential circumstance or event, and typically a function of (1) the adverse impacts that would arise if the circumstance or event occurs; and (2) the likelihood of occurrence

Robust Architecture
Architecture that performs sufficiently well under a variety of possible future developments

Robustness
Dependability with respect to external faults (including malicious external actions)

Role Player
A human that acts according to a given script during the execution of a system and could be replaced in principle by a cyber-physical system

RUI Connecting Strategy
Part of the interface specification of RUIs is the RUI connecting strategy which searches for desired, w.r.t. connections available, and compatible RUIs of other CSs and connects them until they either become undesirable, unavailable, or incompatible

Safety
The absence of catastrophic consequences on the user(s) and on the environment

Sampling
The observation of the value of relevant state variables at selected observation instants

Scenario
A scenario is a projected or imagined sequence of events describing what could possibly happen in the future (or have happened in the past)

Scenario-Based Reasoning (SBR)
Systematic approach to generate, evaluate and manage different scenarios in a given context

Second
An internationally standardized time measurement unit where the duration of a second is defined as 9 192 631 770 periods of oscillation of a specified transition of the Cesium 133 atom

Security
The composition of confidentiality, integrity, and availability; security requires in effect the concurrent existence of availability for authorized actions only, confidentiality, and integrity (with "improper" meaning "unauthorized")

Security Level
Specification of the level of security to be achieved through the establishment and maintenance of protective measures

Security Policy
Given identified subjects and objects, there must be a set of rules that are used by the system to determine whether a given subject can be permitted to gain access to a specific object. This is called the security policy

Semantic Specification
The specification that explains the meaning of the named syntactic units

Send Instant
The instant when the first bit of a message leaves the sender

Sensor
A sensor is an interface device that observes the system environment and produces data (a bit pattern) that can be explained by the design of the sensor and its placement in the physical environment

Service
The intended behavior of a system

Service Composition
The integration of multiple services into a new service is called service composition

Service Consumer
The component that requires a service

Service Discovery
Service discovery is the process where service consumers match their service requirements against the available Interface Service Specifications (S-Specs) in a service registry

Service Level Agreement (SLA)
A SLA defines a set of Service Level Objectives (SLOs), the price of the service, and compensation actions in case of failure to deliver a committed service

Service Level Objective (SLO)
A functional or non-functional objective that can be evaluated by observing the service provider to either achieved or not-achieved. Objectives are based on measurable quality metrics

Service Provider
The component that provides a service

Service Registry
The service registry is a repository of Interface Service Specifications (S-Specs) of service providers

Signal
An event that is used to convey information typically by prearrangement between the parties concerned

Situation assessment
Situation assessment is the process of achieving, acquiring or maintaining situation awareness

Solution Path/Plan
A path of intermediate states from the initial state to the goal state, considering the given constraints

Sparse Events
Events that occur in the active interval of the sparse time

Sparse Time
A time-base in a distributed computer system where the physical time is partitioned into an infinite sequence of active and passive intervals

Sphere of Control (SoC)
The sphere of control of a system during an IoD is defined by the set of entities that are under the control of the system

Stability
The stability of a clock is a measure that denotes the constancy of the oscillator frequency during the IoD

Start Signal
An event that causes the start of an action

State
The state of a system at a given instant is the totality of the information from the past that can have an influence on the future behavior of a system

State Space
The state space of a system is formed by the totality of all possible values of the state variables during the IoD

State Variable
A variable that holds information about the state

Statefull Action
An action that reads, consumes, writes or produces state

Statefull System
A system that contains state at a considered level of abstraction

Stateless Action
An action that produces output on the basis of input only and does not read, consume, write or produce state

Stateless System
A system that does not contain state at a considered level of abstraction

Stigmergic Information Flow
The information flow between a sending CS and a receiving CS where the sending CS initiates a state change in the environment and the receiving CS observes the new state of the environment

Stigmergy
Stigmergy is a mechanism of indirect coordination between agents or actions. The principle is that the trace left in the environment by an action stimulates the performance of a next action, by the same or a different agent

Subject
An active user, process, or device that causes information to flow among objects or changes the system state

Subsystem
A subordinate system that is a part of an encompassing system

Supervenience
The principle of Supervenience states that (Sup i) a given emerging phenomenon at the macro level can emerge out of many different arrangements or interactions of the parts at the micro-level while (Sup ii) a difference in the emerging phenomena at the macro level requires a difference in the arrangements or the interactions of thc parts at the micro level

Symmetric Cryptography
Cryptography using the same key for both encryption and decryption

Symmetric Key
A cryptographic key that is used to perform both encryption and decryption

Syntactic Compatibility
The syntactic chunks sent by the sender are received by the receiver without any modification

Syntactic Specification
The specification that explains how the data field of a message is structured into syntactic units and assigns names to these syntactic units

System
An entity that is capable of interacting with its environment and may be sensitive to the progression of time

System Architecture
The blueprint of a design that establishes the overall structure, the major building blocks and the interactions among these major building blocks and the environment

System Boundary
A dividing line between two systems or between a system and its environment

System Effectiveness
The system's behavior as compared to the desired behavior

System Efficiency
The amount of resources the system needs to act in its environment

System Performance
The combination of system effectiveness and system efficiency

System Resources
Renewable or consumable goods used to achieve a certain goal. E.g., a CPU, CPU-time, electricity

System-of-Systems (SoS)
An SoS is an integration of a finite number of constituent systems (CS) which are independent and operable, and which are networked

together for a period of time to achieve a certain higher goal

Temporal Order
The temporal order of events is the order of events on the timeline

Thing
A physical entity that has an identifiable existence in the physical world

Threat
Any circumstance or event with the potential to adversely impact organizational operations (including mission, functions, image, or reputation), organizational assets, individuals, or other organizations through a system via unauthorized access, destruction, disclosure, modification of information, and/or denial of service

Tick
The event that increments the register is called the tick of the clock

Time
A continuous measureable physical quantity in which events occur in a sequence proceeding from the past to the present to the future

Timeline
A dense line denoting the independent progression of time from the past to the future

Timely Message
A message is timely if it is in agreement with the temporal specification

Timestamp (of an event)
The timestamp of an event is the state of a selected clock at the instant of event occurrence

Time-aware SoS
A SoS is time-aware if its Constituent Systems (CSs) can use a global timebase in order to timely conduct output actions and consistently—within the whole SoS – establish the temporal order of observed events

Time-Synchronization Interface (TSI)
The TSI enables external time-synchronization to establish a global timebase for time-aware SoSs

Time-Triggered (TT) Action
An action where the start signal is derived from the progression of time

Top Down Design
A hierarchical design methodology where the design starts at the top of the holarchy or formal hierarchy

Transaction
A related sequence of computational actions and communication actions

Transaction Activity Interval
The interval between the start signal and the end signal of a transaction

Transducer
An interface device converting data to energy or vice versa. The device can either be a sensor or an actuator

Trans-Ordinal Law
A Law that explains the emergence of the whole and the new phenomena at the macro-level out of the properties and interactions of the parts at the lower adjacent micro-level

Transport Duration
The duration between the send instant and the receive instant

Transport Specification
This part of the interface specification describes all properties of a message that are needed by the communication system to correctly transport a message from the sender to the receiver(s)

Trusted System
A trusted system or component is one whose failure can break the security policy

TT-Message
A TT-Message (Time-Triggered) is an error controlled transport service for the transmission of periodic messages from a sender to many receivers

Unexplained Emergence
An emergent phenomenon that is observed at the macro level is unexplained emergent if, after a careful analysis of the emergent phenomenon, no trans-ordinal law that explains the appearance of the emergent phenomenon at the macro level out of the properties and interactions of the parts at the adjacent micro level is known (at least at present)

Universe of Discourse (UoD)
The Universe of Discourse comprises the set of entities and the relations among the entities that are of interest when modeling the selected view of the world

Unmanaged SoS evolution
Ongoing modification of the SoS that occurs as a result of ongoing changes in (some of) its CSs

Utility Interface
An interface of a CS that is used for the configuration, or the control, or the observation of the behavior of the CS

Valid Message
A message is valid if its checksum and contents are in agreement

Validity Instant
The instant up until an interface specification remains valid and a new, possibly changed interface specification becomes effective

Value
An element of the admissible value set of an attribute

Value Correct Message
A message is value-correct if it is in agreement with the value specification

Variable
A tuple consisting of data and a name, where the name points to the explanation of the data

Virtual SoS
Lack of central purpose and central alignment

Vulnerability
Weakness in a system, system security procedures, internal controls, or implementation that could be exploited by a threat

Wander
The long-term phase variations of the significant instants of a timing signal from their ideal position on the time-line (where long-term implies here that these variation of frequency are less than 10 Hz) (see also jitter)

Worst Case Execution Time (WCET)
The worst-case data independent execution time required to execute an action on a given computer

Author Index

Author Index

Printed in the United States
By Bookmasters